Emerging Technologies in Aquaponics Food Production Systems

Emerging Technologies in Aquaponics Food Production Systems

Editor: Louis Henry

MURPHY & MOORE
www.murphy-moorepublishing.com

Murphy & Moore Publishing,
1 Rockefeller Plaza,
New York City, NY 10020, USA

Visit us on the World Wide Web at:
www.murphy-moorepublishing.com

ISBN: 978-1-63987-187-2 (Hardback)

Cataloging-in-Publication Data

Emerging technologies in aquaponics food production systems /
edited by Louis Henry.
p. cm.
Includes bibliographical references and index.
ISBN 978-1-63987-187-2
1. Aquaponics. 2. Aquaculture. 3. Hydroponics. 4. Food industry and trade--Appropriate technology. I. Henry, Louis.

SB126.5 .E44 2022
631.585--dc23

Table of Contents

Preface

Aquaponics refers to a food production system which makes use of the principles of both aquaculture and hydroponics. It involves feeding nutrient rich aquaculture water to a hydroponically grown plant. The process is accompanied by nitrifying bacteria for conversion of ammonia into nitrates. The components of a typical aquaponics system include a rearing tank, settling basin, biofilter, sump and a hydroponic subsystem. Some common applications of hydroponic systems consist of deep water raft aquaponics, recirculating aquaponics and reciprocating aquaponics. Lettuce, basil, tomatoes and okra are some popular vegetables grown using this technique. This book includes some of the vital pieces of work being conducted across the world, on various topics related to aquaponics. It traces the progress of this field and highlights some of its key concepts and applications. The readers would gain knowledge that would broaden their perspective about emerging technologies in aquaponics food production systems.

This book is a comprehensive compilation of works of different researchers from varied parts of the world. It includes valuable experiences of the researchers with the sole objective of providing the readers (learners) with a proper knowledge of the concerned field. This book will be beneficial in evoking inspiration and enhancing the knowledge of the interested readers.

In the end, I would like to extend my heartiest thanks to the authors who worked with great determination on their chapters. I also appreciate the publisher's support in the course of the book. I would also like to deeply acknowledge my family who stood by me as a source of inspiration during the project.

Editor

1

Regulatory Frameworks for Aquaponics in the European Union

Tilman Reinhardt, Kyra Hoevenaars, and Alyssa Joyce

Abstract This chapter provides an overview of the regulatory framework for aquaponics and the perspectives for European Union (EU) policy. Using Germany as an example, we analyze the specific regulations concerning construction and operation of aquaponic facilities and the commercialization of aquaponic products. We then show how aquaponics fits in with different EU policies and how it might contribute to EU sustainability goals. In the end, we provide some recommendations on how institutional conditions could be improved for aquaponics as an emerging technological innovation system.

Keywords Aquaponics · Law · Regulatory framework · Organic production · Animal welfare · Food labeling · Food safety policies

Introduction

Regulatory frameworks can have a decisive influence on the implementation of sustainable technologies. However, there are currently no specific regulations or policies for aquaponics in the European Union (EU) or most of its member states. One of the reasons perhaps is that it falls at the intersection of various larger fields (industrial aquaculture, wastewater recycling, hydroponics, urban aquaculture), wherein producers are subject to a variety of potentially disparate and conflicting regulations. The following chapter provides an overview of the regulatory

T. Reinhardt (✉)
GFA Consulting Group, Berlin, Germany
e-mail: tilman.reinhardt@gfa-group.de

K. Hoevenaars
AquaBioTech Group, Mosta, Malta
e-mail: kyh@aquabt.com

A. Joyce
Department of Marine Science, University of Gothenburg, Gothenburg, Sweden
e-mail: alyssa.joyce@gu.se

framework for aquaponics and gives some perspectives on how the development of aquaponics could be supported through EU policy. It builds on the work by Koenig et al. (2018) who have analyzed aquaponics through the theoretical framework for emerging technological innovation systems (see Bergek et al. 2008) and have shown how development pathways for this aquaponics might be influenced by institutional conditions.

The first section provides an overview of specific regulations that govern each step in development of aquaponic enterprises, i.e., construction, operation, and commercialization. It analyzes how this regulatory framework provides incentives or disincentives for individual entrepreneurs and market actors to invest in aquaponics. The second section analyzes how aquaponics fits in with different EU policies and how aquaponics can contribute to achieving EU sustainability goals. It then shows how the policies and strategies need to be redefined in order to provide better opportunities in this sector. In the third section, we draw conclusions from lessons learned in the first sections in order to provide policy recommendations.

Note: The first section summarizes the findings of a legal guideline on the feasibility of aquaponics projects in Germany. Detailed references to German regulations and case law have been left out for better readability. A German version with references to specific provisions and relevant case law is available on request from the authors. Parts of the second sub-chapter have been published in Eco cycles. Reference: Hoevenaars, K., Junge, R., Bardocz, T., and Leskovec, M. 2018. EU policies: New opportunities for aquaponics. Eco cycles 4(1): 10–15. DOI: 10.109040/ecocycles.v4il.87.

Legal Framework for Aquaponics

In this first section, our goal is to provide an overview of relevant regulations for the construction and operation of aquaponics facilities and the marketing of aquaponically produced products. We focus specifically on Germany, as it is impossible to extrapolate across the EU given that several important regulations, especially regarding zoning and construction, have not been harmonized across the EU. Although we focus on the German context, similar findings regarding planning law have also been reported in other countries (Joly et al. 2015).

Regulations on Construction

Aquaponic facilities must comply with various planning, building and water regulations, many of which do not fall under EU competence. In Germany, the general framework for planning and water law is harmonized at the national level, while building and local water-use regulations are determined at the state level, with urban and regional planning covered at the municipal level.

Planning Law

Planning law regulates the use of the soil and the area-related requirements for construction projects. There is a major distinction between projects in outlying and inner urban areas.

According to Section 35 of the German Building Code, outlying areas should be kept free of buildings and are reserved for certain uses, such as agriculture or renewable energy production. Whether or not aquaponics constitutes agriculture in that sense remains an unanswered question: while courts have ruled that soilless cultivation of vegetables such as hydroponics can be considered *agriculture*, the case is less clear for aquaculture in indoor facilities with no connection to the natural water cycle. The definition of agriculture in Section 201 of the Building Code only recognizes fisheries. Most courts therefore view recirculatory aquaculture systems as commercial rather than agricultural enterprises. Recently, however, the administrative court of Hamburg has ruled that a plant for fish and crustacean production can be considered agricultural, if the majority of the required feed could theoretically be produced on the agricultural land, belonging to the farm regardless of the type of fish produced, or whether feed is actually produced on the farm. This exception might not be viable however in cases, where agriculturally sourced feed is not used at all. In practice, aquaculture operations were often set up in connection with biogas plants. As farmers received an additional bonus on the feed-in-tariff for *cogenerating* plants (i.e. plants that also produce heat), there was an incentive to install heat-absorbing aquaculture next to the biogas plant.

Additional restrictions may apply in protected areas. The construction of aquaculture facilities is seen as problematic especially adjacent to natural water bodies. Exceptions for agriculture are only available for existing facilities. This has created a number of problems in traditional fishery areas, such as Mecklenburg, where many professional fishermen have an interest and necessary skills to operate ancillary businesses such as aquaculture or aquaponics (Paetsch 2013). Given that aquaponic systems do not rely on the natural water cycle, they could provide a creative possibility for new enterprise if their benefits were assessed and recognized by relevant authorities.

However, regardless of their size, aquaponics facilities do not require an environmental impact assessment, which is only a requirement for fish farms that discharge waste into surface waters.

Urban Areas

Many proponents envision aquaponics as a possibility for urban agriculture, given that commercial facilities can be built on rooftops or unused warehouses to allow for direct delivery of products to supermarkets in urban centers. Semicommercial systems can also be located in residential areas (*backyard aquaponics*). Under German planning law, the permitting of a facility depends on its classification and the area where it is located. Commercial aquaponics farms can be classified as either commercial or horticultural businesses. As such, they are generally not allowed in residential areas. In villages and mixed-use areas, both commercial and horticultural enterprises are allowed. In commercial and industrial areas, only commercial, but not horticultural, businesses are possible.

As aquaponic facilities have comparatively few noise and odor issues, they may be allowed on an exceptional basis even in areas where they are not currently admissible under planning laws. However, obtaining an exception creates additional administrative burdens and uncertainty, which could present an obstacle to scaling-up of the technology. Project-specific planning allows for cooperation with planning authorities but, in practice, is only relevant for large-scale projects due to the costs involved.

Backyard aquaponics plants might be allowed in all zones under the exception for *ancillary facilities for keeping of small animals.* However, ancillary facilities must be noncommercial and are interpreted differently by different district authorities. Some municipalities take a rather restrictive approach and only allow traditional forms of small animals such as dogs, chicken, pigeons, etc.

Building Law

The structural-technical requirements for buildings and administrative procedures for obtaining building permits are regulated by state-level building regulations, and while building codes follow, a so-called model building code, there can be substantial differences between states.

All construction products must comply with EU Regulation 305/2011, which requires a Declaration of Conformity to technical standards. For small wastewater treatment plants, the technical standard EN 12566 CEN applies. Rooftop systems may require special facilities for fire protection and affect minimum clearances. The static stability of the building must not be affected.

Although some of the components of an aquaponics facility, especially green houses or water tanks, do not need an individual building permit, the installation of a commercial food production system will usually require a building permit, particularly if the building has served a different purpose before. The procedure to obtain this permit might constitute a significant administrative and financial hurdle. Once obtained, however, it can also be seen as providing increased stability for outside investors given regulations will be viewed as having been met.

Water Law

Aquaponic systems do not necessarily depend on the use of surface water. Ideally water leaves an aquaponic system only via evapotranspiration or as water retained in the vegetables produced. We would argue that such facilities therefore should not require a permit under the water act or wastewater regulations. This could provide a major regulatory advantage compared, for instance, to traditional aquaculture or aquaculture, for which increasingly restrictive water and wastewater regulations constitute a significant obstacle to new enterprises. Savings on wastewater discharge fees present an incentive for implementing such systems.

However, it is not totally certain, if courts would follow this line of argument. Water might be considered wastewater, the moment it has been used for aquaculture. It would then be subject to the rules on wastewater disposal, which generally require disposal through centralized facilities. For example, the Higher Administrative Court of Berlin recently disallowed the use of a reed bed to clear domestic grey and blackwater, where reed was later used for energy use. The court explicitly stated that no right to multiple use of the water exists under German Water Law. In this case, a special permit for decentralized wastewater disposal would be required, and the entire aquaponics installation would have to comply with rules on wastewater disposal facilities.

The only real waste produced in aquaponic systems is filter sludge (which can be avoided, if an additional cycle for remineralization of this sludge is integrated or sludge is degraded on-site, e.g., through vermicomposting: if the filter sludge can be used on-site, no registration under the fertilizer law is required (see below)). If used outside the premises, regulations related to organic waste or sewage sludge disposal (more restrictive) apply. Filter sludge would be considered sewage sludge if the aquaponic system as a whole is considered a *wastewater treatment plant:* whether this applies in practice remains to be determined.

Conclusions on Construction

Fish farms present few problems in regard to noise and odor. One might therefore assume that aquaponic systems could be permitted with more ease than other animal production facilities. However, aquaponics does not fit well into the German legal framework.

As aquaponic production does not depend on the use of soil, installations may not be "agricultural enough" for outlying areas, i.e., agricultural lands. On the other hand, aquaponics might be "too agricultural" for urban areas, as urban agriculture is not considered a relevant category under German planning law. In particular, aquaponics may be generally inadmissible in commercial, industrial, and residential areas.

Commercial aquaponics facilities always require a building permit even if they are installed in pre-existing buildings which themselves do not require new building permits.

Aquaponics pioneers with very visible urban projects such as ECF or Urban Farmers seem to have coped well with the existing regulatory framework. However, planning law issues could present a relevant problem for scaling up the technology, in which case projects need to be developed in close consultation with the authorities in order to avoid future conflicts and to provide certainty for investors.

A major regulatory advantage of aquaponics may lie in the fact that little or no wastewater is produced, thus reducing the need for wastewater removal. Wastewater permits and fees have been reported to be significant obstacles for conventional fish farmers. As wastewater fees will probably be calculated according to pollution loading in the future, they may form an even stronger incentive to think about alternative types of wastewater disposal in the future (Schendel 2016). However as the water law generally does not provide for multiple uses, a legal clarification would be very important to create certainty for producers.

Apart from this, regulatory conditions in the German water sector do not particularly favor innovation. German water law strictly adheres to the paradigm of centralized sewage and generally does not allow for decentralized recycling of material flows and other forms of "creative ecology." Unlike in the waste sector, where the regulatory framework has given strong incentives to the private sector to consider waste as a resource, regulation of wastewater sector does not create incentives for the private sector to create and implement innovative recycling technologies.

Regulations on Aquaponic Production

Aquaponic production is subject to regulations for plant and for animal production at all stages of production and processing. Under the regulatory approach "from farm to fork," many relevant regulations have been harmonized at the European level (especially through the so-called EU hygiene package). However, a few exemptions exist for small producers selling directly to customers.

Hydroponic Production

Hydroponic production is subject to comparatively few regulations: growing media need an EU approval. Using fish wastes as fertilizer requires no authorization under German fertilizer laws if those fish wastes are derived from aquaculture.

The most significant restrictions concern the use of pesticides (note: in single-cycle aquaponic systems, pesticide use is inherently limited due to the pesticides' toxicity for fish; however, pesticide use is possible in multi-cycle (decoupled) aquaponic systems where water does not return from the plants to the fish components). The German Plant Protection Act generally imposes *integrated pest management* which means that preventive measures and the promotion of natural response mechanisms (e.g., suitable locations, substrates, varieties, seeds and fertilizers, as well as physical and biological control measures) have to be given priority before pesticide use. The use of invasive species for biological pest control is prohibited.

Pesticides may only be used by qualified personnel. Only pesticides that are approved under European regulation (EC) 1107/2009 may be used. Regulation (EC) 1107/2009 also includes rules on the use, storage, and disposal of pesticides.

Before harvesting, certain waiting periods must be observed. Residues in vegetables may not exceed certain *maximum residue levels* (MRLs). A free online database of MRLs is provided by the Directorate General for Health and Consumers (DG SANCO).

Aquaculture

In contrast to horticulture, aquaculture is carefully regulated through many different regulations. However a distinct *aquaculture law* neither exists at the national nor European level, and state-level fishery laws only regulate fishing in natural water bodies.

Cultivation of Non-native Species

The most commonly cultivated fish in aquaponics systems are tropical species such as tilapia or African catfish. However, the complex rules of Regulation (EC) 708/2007 concerning the use of alien species in aquaculture generally do not apply to closed recirculating aquaculture facilities (registered in a directory of recirculating aquaculture facilities). Some countries (e.g., Spain and Portugal, but not Germany or France), however, have decided to ban some types of exotic fish outright, which also affects the possibilities of cultivating them in closed facilities.

Regulation on Fish Diseases in Aquaculture

All aquaculture producers are subject to German regulation on fish diseases, which implements European Directive 2006/88/EC on animal health requirements for aquaculture animals and products thereof and on the prevention and control of certain diseases in aquatic animals (Ministry of Agriculture of Bavaria 2010). Under this regulation, aquaculture operations generally require permits by the local veterinary authorities. However, producers who only sell small quantities of fish directly to consumers or to local retailers only need to register certain information such as name and address, location and size of the operation, source of water supply, amount of fish held, and fish species.

Most importantly, the regulation on fish diseases imposes a duty on operators of fish farms to inform the local veterinary authorities in the event of a suspected disease outbreak. Veterinary authorities may then implement necessary control measures, which in some cases may involve destroying the entire stock should there be concern of disease spread.

Note: European animal health law, which was previously regulated relatively confusing in some 400 individual acts, is unified under Regulation (EU) 2016/429. However, the regulation only enters into force only on 21 April 2021. The content concerning fish diseases will not change (Art. 173 et seq. Regulation (EU) 2016/429).

Regulations on Animal Feed

The ability to source sustainable animal feeds is a key prerequisite for sustainable food production. In comparison to land animals, fish have a much better feed conversion rate; however, many higher trophic-level fish species require a certain portion of their feed to be derived from protein and fat from animal origin (e.g., fish meal). Feeding insects or insect larvae to fish is often seen as a possible way to increase the sustainability of aquaculture. Insects can be cultivated using waste organic nutrients, in some cases derived from animal wastes, including offal.

Animals for human consumption however must not be fed proteins of animal origin (with the exception of fish protein) according to Regulation (EC) 999/2001, which was implemented in reaction to BSE crisis in the 1990s. While it is sometimes argued that the ban of feed sources from animal protein should not apply to insects, which were not considered as potential feed sources in 2001, the use of insect feed is disallowed in practice by the German veterinary authorities.

Currently, some pet food is already produced using insect proteins (e.g., dog food from Brandenburg-based start-up Tenetrio). Given the increased interest in using insect protein for animal feed, several legislative changes have been made at the European level to allow for the feeding of insect protein based on adaptations to the existing regulatory frameworks (Smith and Pryor 2015). Since 2017, a so-called risk profile of the European Food Safety Authority (EFSA) is available (EFSA Journal 2015; 13 (10): 4257). Insects may be allowed as feed in aquaculture from 2018. However, some restrictions remain in place: in particular insects to be used as animal feed must not have fed on human or ruminant waste products. Insect production also poses some addition open regulatory questions, e.g., welfare issues regarding standard procedures for killing.

Regulation on Animal Welfare

Compared to other livestock, there are far few animal welfare regulations (Chapter) for handling and killing of fish. Although it is generally accepted that fish can feel pain, there is a lack of scientific evidence to justify restrictions for animal welfare (Studer/Kalkınç 2001). On the European level, there are only a few non-binding recommendations initiated in 2006 by the European Commission. According to Article 22 of these recommendations, a revised version based on new scientific evidence was anticipated by 2011, but to date EFSA has only published species-specific recommendations for certain types of fish, as well as special provisions on the transport of fish. Article 25 lit f – h. and Annex XIII of Regulation (EC) 889/2008 on organic production and labeling of organic products also contains species-specific rules on stocking density. As organic labeling is not available for fish from recirculatory aquaculture (see below), these rules are not relevant for aquaponics. Most private certification standards also do not consider animal welfare aspects (Stamer 2009).

Under Section 11 of the German law on animal welfare, keeping animals for commercial purposes generally requires a permit. To obtain this permit, one must demonstrate appropriate training or previous professional experience in animal husbandry and show that the production system provides adequate nutritional and housing facilities (Windstoßer 2011). Operations are considered commercial when expected sales exceed € 2000 per year.

According to Section 11 para 1 no. 8 TierSchG, no permit is needed for the commercial keeping of "farm animals." Whether fish can be considered *farm animals* in this sense is unclear. Exceptions to the animal welfare law are generally interpreted strictly: species are only considered *farm animals* if the necessary skills for keeping them can be acquired anytime, anywhere and there exists sufficient experience regarding the keeping of a species (Windstoßer 2011). This may not be the case for some types of tropical fish that differ fundamentally from native species (e.g., *Arapaima*, whose use in aquaculture is currently being explored at IGB Berlin).

The Administrative Court of Cologne has recently examined animal welfare aspects when ruling on the admissibility of a so-called fish spa, where Kangal fish were kept with the goal of using them to clean human feet. Operators of this fish spa were able to prove through veterinary reports that animal welfare was not being compromised and as such, a permit was granted.

Regulation on Fish Slaughter

Animal slaughter is regulated by European Regulation (EU) 1099/2009 as well as the German Decree of 20.12. 2012 (Federal Law Gazette I, p. 2982).

According to Recital 11 of Regulation (EU) 1099/2009, fish are physiologically different from terrestrial animals, and thus farmed fish can be slaughtered and killed with fewer animal welfare restrictions, in this case with specific implications for the inspection process. Furthermore, research on the stunning of fish is far less developed than for other farmed species. Separate standards should be established on the protection of fish at killing. Therefore, provisions applicable to fish should, at present, be limited to the key principle.

Under the general rule of Art. 3 Section 1 of Regulation (EU) 1099/2009, animals shall be spared any avoidable pain, distress, or suffering during their killing and related operations. However, there is no explicit obligation to stun fish before slaughter. That said, EU member states may maintain or adopt national rules aimed at ensuring more extensive protection of animals at the time of killing than those contained in this Regulation (Art. 26).

For instance, in Germany, slaughtering fish is subject to stricter conditions than those imposed by Regulation (EU) 1099/2009: as such, all types of fish except flatfish and eel have to be stunned before killing. Those doing the killing need a certificate of competence. Appropriate stunning methods may differ from species to species, with implications for producers: in Switzerland, aquaculture producer reportedly had to close his operation as the local veterinary authorities did not allow stunning using the ice water method, which he was employing. Defining appropriate killing methods for different fish species is the subject of an ongoing research project funded by BLE at the University of Veterinary Medicine Hannover and may lead to more restrictive regulations in the future.

Animal welfare aspects may also restrict certain forms of marketing and selling of fish. For example, the Higher Administrative Court of Bremen has forbidden placing farmed fish into ponds, from which were to be fished by recreational anglers as this was considered "unnecessarily harmful."

Regulation on Vocational Training of Fish Farmers

The German federal regulation on the vocational training of fish farmers does not mention aquaponics. Some private companies offer courses on aquaponics on the German market. However, it is not clear whether such courses are considered sufficient to obtain necessary permits (e.g., for pesticide use, commercial keeping of animals, slaughtering, etc.).

Hygiene Law

Hygiene law is harmonized on the European level through Regulations (EC) 852/2004, 853/2004, and 854/2004.

As a general rule, all food business operators, regardless of product, have to comply with EU hygiene law. As such, they must comply with the general standards of hygiene and management in Annex I and II of Regulation 852/2004, including basic requirements on production processes and personal hygiene, as well as appropriate waste treatment. They have to keep a register of the origin of animal feeds, as well as the use of pesticides and veterinary drugs. Measures to avoid risks must be documented in an appropriate manner.

According to Annex II Chapter of Regulation (EC) 852/2004, food has to be protected against any contamination at all stages of production, processing, and distribution. Under Regulation (EC) No 178/2002, contamination may refer to any biological, chemical, or physical agent in a food or a condition of a foodstuff, which can cause adverse health effects. Food business operators must implement and maintain a HACCP (hazard analysis and critical control points) system, which has to be certified by accredited certification bodies. Details are agreed upon with local authorities.

EU hygiene legislation does not apply to the direct supply of small quantities of primary products to the final consumer or to local retail establishments. Small amounts are defined as household amounts for direct delivery to consumers or for local retail as for usual daily consumption. Primary production in the case of fish includes catching, slaughtering, bleeding, heading, gutting, removing fins, refrigeration, and wrapping. Activities such as flash-freezing, filleting, vacuum packing, or smoking will result in the fish no longer being considered primary local production.

Under German law certain restrictions on food hygiene exist however even for direct local food suppliers (Annex 1 LMHV).

Registration or authorization requirements in Regulation (EC) No 853/2004 depend on volume and type of processing. No registration or authorization is required for the supply of primary products in household quantities directly at the place of production, processing, or storage (including nearby markets). It is also possible to supply retail establishments (supermarkets, restaurants), consumers, or restaurants within a radius of 100 km. If primary products are delivered to end consumers, or to restaurants in larger quantities, the company must register and demonstrate their ability to meet food hygiene requirements. If more than one-third of the animal-derived products are sold to retail outlets outside the region (radius of 100 km), a public health license is also required.

Conclusions on Production

The legal requirements for aquaponic production are not higher than for the production of fish or vegetables. However, the large number of applicable laws reflects the complexity of aquaponics.

Compared to livestock farming, aquaculture may appear less regulated, especially in the area of animal welfare law. However, legal "gray areas" and the corresponding uncertainty are not always to the advantage of producers. Without established administrative practices, there is a considerable risk of conflicts (c.f. for the cited case in Switzerland, where an aquaculture operation was shut down, because the producer was not allowed to kill the fish in specific ways). In addition, the large number of applicable regulations can be burdensome, especially where European and national regulations coexist (e.g., on animal welfare or hygiene). As there is no harmonized law on aquaculture in Germany, producers need various permits from different authorities. Authorities often have little experience with nontraditional aquaculture, and as such, uncertain administrative requirements can be discouraging for entrepreneurs. Given the relative newness of commercial aquaponics, potential producers are strongly advised to contact local authorities at an early stage. In the case of larger, commercial plants, operators should probably contact veterinary and hygiene authorities prior to beginning construction.

The increasingly strict requirements of European hygiene laws can also constitute a significant burden, especially for small enterprises who wish to market directly to consumers or local restaurants (Schulz et al. 2013). However, whether exemptions for direct sellers are of practical help to aquaponic operators remains to be determined. The few existing aquaponics facilities in Germany currently demonstrate the necessity for producers to depend on a variety of sales channels and the need to create various forms of ancillary revenue (guided tours, secondary processing, cooking classes, etc.). Exemptions for direct selling therefore may become irrelevant if hygiene standards have to be met for other reasons.

Commercialization

Commercialization of aquaponic products is affected by different regulatory regimes. Hygiene regulations not only concern food production but also food retailing (see above). Business and tax laws, labeling regulations, or special certifications, such as the EU organic labeling regulations, may also be relevant.

Business and Tax Law

Agriculture is privileged in a number of ways under German business law: the marketing of self-produced agricultural products through farm stores, from the field or from a market stall, is not considered a business under German law and therefore requires no registration. This exception extends to the first stage of processing, i.e., cleaning and gutting, filleting and smoking in the case of fish, or in the case of fruits and vegetables, peeling, chopping, cooking, as well as the production of juices and wine (Chamber of Agriculture Rhineland-Palatinate 2015). Direct selling of agricultural products is also exempt from legal restrictions on opening times and the Sunday sales ban. Considering the low costs and low administrative requirements of a business registration, however, these privileges may not constitute a relevant advantage.

Tax privileges for agricultural production may be of greater practical importance. Regardless of size, aquaculture operations are subject to so-called average-rate VAT taxation, which offers a greatly reduced effective rate of value-added-tax that enables local producers to sell at more competitive prices relative to international imports.

In the income tax code, there are considerable privileges for "small farms" (i.e., turnover <500,000 €, farm size <20 hectares without special uses). If certain area limits are respected (600 m^2 of under-glass vegetables, 1600m^2 of ponds) income from aquaculture and vegetable growing is not taxed at all; even if the cultivated areas exceed these limits, effective tax rates are extremely low. As a result, the operation of aquaponics for small farmers might essentially be considered tax-free.

Regulations on Food Labeling

Rules on food labeling have largely been harmonized at the European level through European Regulation 1169/2011 on food packaging. Besides the formal rules, however, voluntary labeling standards play an even bigger role in the marketplace (see Sodano et al. 2008). In the case of the EU organic label, the voluntary standard is also regulated by law. In other cases, the rules of private certification schemes have to be followed.

General Labeling Rules

The general rules for selling packaged products are laid down in European Regulation 1169/2011 (e.g., the duty to include a list of ingredients, etc.). As a general rule, Art. 7 para. 1 Regulation (EU) 1169/2011 bans misleading claims on food packaging.

In addition to these general rules, Regulation (EU) 1379/2013 contains special rules on consumer information regarding aquaculture products. For example, under Art. 38 of Regulation (EU) 1379/2013, the member state or the third country, in which the aquaculture product has acquired more than half of its final weight, must be correctly named on the label.

EU Organic Regulation

Products from aquaponics are not eligible for labeling as organic under the current EU regulations for organic products. Article 4, Regulation (EC) 889/2008, explicitly forbids the use of hydroponics in organic farming. Recital 4 states that organic/biological crop production is based on the principle that plants obtain their food primarily from the soil ecosystem. For aquaculture products, Art. 25 g Regulation (EC) 710/2009 forbids the use of closed-circuit systems, and according to Recital 11 of Regulation (EC) 710/2009, this follows from the principle that organic production should be as close to nature as possible. These rules will not change in the new EU organic labeling regulation adopted in 2018, which will enter into force in 2021.

Laws preventing organic certification of hydroponic products are not shared by countries such as the USA and Australia, where hydroponic/aquaponic products can be certified organic.

Private Labels

Although there are currently no specific certification schemes for aquaponics, a number of certifications are available for aquaculture. Private certifications are usually "awarded" for a certain period of time, if private certain certification bodies can verify that productions comply with the criteria defined by their labeling standard. Private labeling schemes are usually purely contractual, with standards set by private institutions, which are subject only to general legal obligations (e.g., antitrust law). While the design of certification systems is increasingly in the purview of European legislation, including certification for aquaculture (cf. Commission report on options with regard to the allocation of EU eco-label for fishery and aquaculture products from 05.18.2016, COM see. (2016) 263 final), no concrete obligations for labeling schemes for aquaculture exist to date under European law.

Labeling schemes are essential, especially in business-to-business (B2B) relationships between producers/processors and retailers. Increasingly, however, certification also plays a role in marketing to final consumers (B2C). Beside qualitative aspects, B2C often certify compliance with certain environmental and social standards. Certification schemes can vary greatly in standards, verification modalities, and cost.

Among existing certification schemes, the Aquaculture Stewardship Council (ASC) may be most relevant for aquaponics producers. This certification has been awarded by the Aquaculture Stewardship Council (ASC) since 2010, as a complement to the better-known Marine Stewardship Council (MSC) program. The ASC is a formally independent, private, nonprofit organization initiated by the WWF and responsible for developing quality, ethical, and sustainability standards with scientific input. Private certification companies (e.g., in Germany TÜV NORD) are accredited by the ASC to confirm compliance with these standards. ASC standards currently exist for the following species: abalone, trout, shrimp, salmon, clams, catfish, and tilapia. Standards, audit manuals, and audit preparation checklists are freely available on the ASC website, and standardized procedures have also been made public. The ASC standard has different priorities, for instance, for the EU organic label (e.g., GMO-based animal feed is not prohibited under ASC).

Aquaculture certifications are also offered by the Cologne-based quality assurance and certification system GLOBALG.A.P. While GLOBALG.A.P. usually focuses on B2B certification for quality assurance in the food retail sector, a consumer label called GGN is also awarded for farmed fish and is often used, for instance, for fish that are not eligible for organic certification because they are wild caught or produced in aquaponics. Consumers can access information via www.my-fish.info.

Market Organizations

Regulation (EU) 1379/2013 on the common organization of markets in fishery and aquaculture products contains detailed rules on the establishment, recognition, objectives, and actions of professional organizations, i.e., producer organizations (Art. 6 ff.) and inter-branch organizations (Art. 11 ff.).

According to Article 8 (3) of Regulation (EU) 1379/2013, producer organizations in the field of aquaculture may, inter alia, make use of the following measures: promoting sustainable aquaculture, notably in terms of environmental protection, animal health, and animal welfare; collecting information on marketed products, including economic information on sales and production forecasts; collecting environmental information; planning the management of aquaculture activities of their members; and supporting programs for professionals to promote sustainable aquaculture products. According to Article 15 of Regulation 1379/2013, producers' organizations may also receive financial support from the EU's maritime and fisheries policies.

Measures taken by inter-branch organizations include, for example, the promotion of aquaculture products of the European Union in a nondiscriminatory manner by using, for example, certification and designations of origin, quality seals, geographical designations, traditional specialties guaranteed, and sustainability merits (Art. 13 lit a Regulation 1379/2013). Inter-branch organizations may adopt rules for the production and marketing of aquaculture products which are stricter than the provisions of European Union or national law (Art. 13 lit c Regulation 1379/2013).

Recognition as a producer organization can have far-reaching consequences; member states may, under certain conditions, make the rules agreed upon within a producer organization binding on all producers in the area (Art. 22 Regulation 1379/2013). Also, agreements, decisions, or concerted practices agreed upon within an inter-branch organization can be made binding on other operators (Article 23 of Regulation 1379/2013). According to Art. 41 VO 1379/2013, producer organizations are largely exempt from antitrust law.

So far, there are no professional associations in the field of aquaponics in the European Association of Fish Producers Organizations (EAPO). However, in 2018 an association for aquaponics has been founded in Germany (http://bundesverband-aquaponik.de/), and an EU Aquaponics Association (EUAA) based in Vienna has been founded at the initiative of several stakeholders assembled in the EU COST Action.

Conclusions on Commercialization

In terms of business and tax law, there are various privileges that could theoretically be exploited by aquaponics operators. Tax benefits could be especially interesting for outside investors. However, it remains to be seen whether certain conditions regarding the legal form of an enterprise and required investment volumes prevent operators from claiming these benefits. So far, the best-known urban aquaponics projects in Germany have not been profitable, so the issue of paying taxes has not arisen.

Under business and tax laws, the thresholds for privileges of small-scale installations and direct marketers are not congruent with the thresholds under hygiene law. A detailed review of operating and marketing concepts is therefore required in each individual case.

The EU organic label is currently out of the question for aquaponic products. However, there are increasingly private opportunities for certification.

Aquaponics and EU Policies

National policies can only be analyzed for each individual country. We therefore concentrate on relevant EU policies.

Overview of Relevant EU Policies

The Common Fisheries Policy (CFP) and the Common Agricultural Policy (CAP) apply to the aquaculture and hydroponics components of aquaponics, respectively (European Commission 2012, European Commission 2013). Policies on food safety, animal health and welfare, plant health, and the environment (waste and water) also apply.

Common Agricultural Policy

The Rural Development Policy, also referred to as the second pillar of CAP, focuses on increasing competitiveness and promoting innovation (Ragonnaud 2017). Each member state has at least one rural development program. Most countries have set goals to provide training, restructure and modernize existing farms, set up new farms, and reduce emissions. Measures against excessive use of inorganic fertilizers were introduced in the CAP as well as environmental policies and are regulated through the EU's Nitrates Directive (Directive 91/676/EEC 1991) and the Water Framework Directive (WFD).

Common Fisheries Policy

The CFP reform and strategic guidelines for the sustainable development of EU aquaculture were issued by the Commission to assist EU countries and stakeholders with challenges that the sector is facing. The emphasis is on facilitating implementation of the Water Framework Directive as it relates to aquaculture (European Commission 2013).

The CFP requires the development of a multiannual national strategic plan in each member state with strategies to promote and develop the aquaculture sector (European Commission 2016). Taking into account their different histories and cultivated species, each member state can support their existing aquaculture technologies but also develop new ones, such as aquaponics. This strategy should lead to an increase in production and reductions in dependence on imports. The main actions planned by member states are simplification of administrative procedures, coordinated spatial planning, enhancement of competitiveness, and promotion of research and development.

In the framework of the CFP, an Aquaculture Advisory Council (AAC) has been established. The main objective of the AAC is to provide advice and recommendations to European institutions and member states on issues related to sustainable development of the aquaculture sector (Sheil 2013).

A goal of both CFP and CAP is to increase competitiveness and sustainability of aquaculture and agriculture, respectively (Massot 2017). One of the objectives in the CFP is to exploit competitive advantage by obtaining high-quality, health, and environmental production standards.

EU Food Safety Policy

The goal of the food safety policy of the EU is to ensure safe and nutritious food from healthy animals and plants while supporting the food industry (European Commission 2014). The integrated food safety policy also includes animal welfare and plant health. In the strategy for animal welfare, there is an action on the welfare of farmed fish; however, there are no specific rules in place (European Commission 2012).

Environmental Policies

Environmental impacts of aquaculture are regulated under a range of EU legal requirements including water quality, biodiversity, and pollution. Environmental policies relevant for aquaponic operators are the strategy on the prevention and recycling of waste (European Commission 2011) and the seventh Environment Action Program (EAP) under the EU Environmental Policy (European Union 2014).

How Aquaponics Can Contribute to Goals in EU Policies and Strategies

Aquaponics can contribute to the development goals mentioned in these policies, with the main factors being reductions of water use and waste from fish production through nutrient recycling. Discharged water is converted into a resource and solid wastes can be upgraded as plant fertilizers. Because modern aquaponics is based on recirculating aquaculture systems, these operations are relatively independent of their location and can contribute to regional food production and value chains even in urban areas. Open aquaculture systems have constraints: water resource use, pollution, localized reduction in benthic biodiversity, significant dredging of water bodies, physical modification of land, changes in water flow, and introduction of alien species (European Union 2016). However, mitigation of most constraints is possible in aquaponic systems. Compared to hydroponic systems, aquaponics reduces the use of mineral, often unsustainably mined, fertilizers.

One of the priorities in the strategic guidelines on aquaculture is to improve access to space and water (European Commission 2013). Competition among different stakeholders and often strict environmental rules limit the further development of open aquaculture systems inside the EU. However, aquaponic systems can be located almost anywhere, including deserts and degraded soil and salty, sandy islands, since a closed-loop uses a minimum of water. Therefore, it can utilize space that is not suitable for other food production systems, like rooftops, abandoned industrial sites, and generally nonarable or contaminated land. Since aquaponics reuses 90–95% of the water, it relies much less on water availability compared to other systems like open aquaculture, hydroponics, and irrigation agriculture.

Just like in recirculating aquaculture systems, a benefit of larger commercial aquaponics systems is the possibility to obtain a high level of biosecurity, in which environmental conditions can be fully controlled ensuring a healthy environment for the fish (Badiola et al. 2012), thus minimizing the risk for diseases and parasite outbreaks (Yanong and Erlacher-Reid 2012). Because of the higher control on production, risk of losses is lower (Yanong and Erlacher-Reid 2012), which can provide aquaponic farmers with a competitive advantage over traditional farmers. On the other hand, using one nitrogen source to culture two products (Somerville et al. 2014) increases the investment risk as both fish and plant production must be maximized in order to make a profit. However, if this is done successfully, combined with the positive perceptions in Western markets of more environmentally friendly products, high revenues can be achieved (Somerville et al. 2014).

An objective in the strategy on the prevention and recycling of waste (European Commission 2011) includes introducing life-cycle thinking that considers a range of environmental impacts. It mentions that the prevention of waste is the priority, followed by reuse, recycling, recovery, and last disposal. Also, one of the priority areas in the seventh EAP targets transformation of the EU into a resource-efficient, low-carbon economy with a special focus on using waste as a resource (European Union 2014). Aquaponics systems minimize waste output (Goddek et al. 2015). The water in aquaponics systems is recirculated, thus wastewater is minimized. By using the fish process water for plant nutrition, organic waste from aquaculture is reused in the hydroponics component of the aquaponics system. The solid waste produced in an aquaponics system can be mineralized and returned to the system or utilized as compost for soil-based agriculture. Aquaponics also promotes local food production, thereby minimizing transport costs. Lastly, placing aquaponics farms in urban settings, it can provide ecological value in cities and play a role in adaptation to climate change.

Financial Support by the EU

The Seventh Framework Program (under the Multiannual Financial Framework of the European Commission) funded a couple of projects related to aquaponics. The EU Framework Program Horizon 2020 (challenge 2 "Food security, sustainable agriculture and forestry, marine and maritime and inland water research, and the bioeconomy" and challenge 5 "Climate action, environment, resource efficiency and raw materials") provides funding to several aquaponics initiatives including COST (European Cooperation in Science and Technology) Action FA1305 "The EU Aquaponics Hub: Realising Sustainable Integrated Fish and Vegetable Production for the EU" to promote innovation and capacity building by a network of researchers and commercial aquaponics companies.

Other possible funding opportunities for aquaponic development projects under the Multiannual Financial Framework of the European Commission include the European Innovation Partnership *Agricultural Productivity and Sustainability*

(EIP-AGRI), *a long term EU-Africa research and innovation partnership on food and nutrition security and sustainable agriculture* (LEAP-AGRI), the European Innovation Council pilot *Small and Medium-Sized Enterprises Instrument* (SME Instrument), the ERANET MED initiative *Partnership on Research and Innovation in the Mediterranean Area* (PRIMA), and the *European Maritime and Fisheries Fund* (EMFF). The EMFF can support research institutions and universities as well as companies; however, it requires different rates of co-funding.

Conclusions on the Overall EU Policy Landscape

None of the EU policies and guidelines so far explicitly mentions aquaponics. According to DG MARE, regulations on aquaponics need to be resolved within the individual member states (COST Action FA1305 2017), e.g., involving action resulting from the respective national strategic plans. Even though there is no explicit EU framework for aquaponics, it is an innovative agricultural system that can contribute to many priorities set through EU policies and strategies. EU support through financial measures is assisting further development of the technology. However, this mostly targets research projects, while the sector also needs assistance for commercial development through support of proof-of-concept projects. As a matter of fact, there are so far very few successful commercial aquaponics systems operating in the EU, so currently, there might not be a necessity for an aquaponics policy. However, ultimately recognizing and covering the technology in existing policies will be beneficial for the development of the sector.

Overall Conclusions and Policy Recommendations

Aquaponics is not only at the nexus of different technologies but also at the nexus of different regulatory and policy fields. While it may provide solutions to various sustainability goals, it seems to fall in the cracks between established legal and political categories. To add to the complexity, the development of aquaponics is affected by regulation from different levels of government. For example, facilitation of urban agriculture has to come from the national or even subnational level, as the EU has no competence in planning law. Major regulatory incentives for the implementation of aquaponic technology could probably be set in water law, which falls under national and EU competence. Implementation of aquaponics could gain significant traction, if aquaculture operations had the obligation or at least financial incentives to deal with wastewater themselves. However this would require a major change in the current regulatory approach.

In the theory on technological innovation systems (TIS), an “institutional alignment” in the formative phase of a TIS is seen as critical. Only if institutions are sufficiently aligned will markets form and provide space for entrepreneurial experimentation to determine commercially viable paths for implementing the technology (Bergek et al. 2008). For institutional alignment to take place, the proponents of the new technology need to be sufficiently organized to contribute to a process of "legitimation" of their technology (Koenig et al. 2018).

As a first step, we therefore recommend that proponents reach out, create, and strengthen links to various stakeholders and among themselves in the relevant professional communities in order to build a case for making aquaponics a legitimate activity. The newly founded EU Aquaponics Association could play a large role in this process. A further step could be the development of certifiable standards in cooperation with established certification systems. With the current lack of a coherent legal framework, such standards would give producers, consumers, administrators, and other relevant parties (e.g., outside investors or insurers, who doubt the safety of aquaponic products) a framework for understanding quality and risks. Such standards could be adapted flexibly to the practical demands of producers. Eventually formal regulations could build on such standards, as they do in other regulatory domains.

On the European level, stakeholders should push for greater recognition of the potential benefits of aquaponics in different policy areas. The EU must provide critical financial support, since commercial implementation of aquaponics is still in its infancy. The EU should also provide a forum for the exchange of best practices for regulatory issues such as construction and wastewater that fall under the competence of member states.

On the national level, stakeholders have to push for a coherent and accessible regulatory framework that is adapted to the realities of modern aquaculture and sets incentives for "creative ecology." Significant advances might even be realized on a subnational level, where political resistance may be overcome more easily. Further research should concentrate on the regulatory strategies of different countries to identify best practices.

References

Badiola M, Mendiola D, Bostock J (2012) Recirculating Aquaculture Systems (RAS) analysis: main issues on management and future challenges. Aquac Eng 51:26–35. https://doi.org/10.1016/j.aquaeng.2012.07.004

Bergek A, Jacobsson S, Carlsson B, Lindmark S, Rickne A (2008) Analyzing the functional dynamics of technological innovation systems: a scheme of analysis. Res Policy 3(37):407–429

COST Action FA1305 (2017) Report on a Workshop with representatives from DG MARE, DG AGRI and DG RTD. 5 p. Available online: http://euaquaponicshub.com/hub/wp-content/uploads/2017/07/Workshop-notes.pdf

Directive 91/676/EEC (1991) Council Directive of 12 December 1991 concerning the protection of waters against pollution caused by nitrates from agricultural sources. Off J Eur Communities 8 p. Available online: http://eur-lex.europa.eu/legal-content/EN/TXT/?uri=celex:31991L0676

European Commission (2011) Report from the commission to the European parliament, the council, the European economic and social committee and the committee of the regions on the Thematic Strategy on the Prevention and Recycling of Waste. 11 p

European Commission (2012) Communication from the commission to the European Parliament, the council, the European Economic and social committee and the committee on the European Union Strategy for the Protection and Welfare of Animals 2012–2015. 12 p. Available online: https://ec.europa.eu/food/sites/food/files/animals/docs/aw_eu_strategy_19012012_en.pdf

European Commission (2013) Communication from the commission to the European Parliament, the council, the European Economic and social committee and the committee of the regions; Strategic guidelines for the sustainable development of EU aquaculture. 12 p. Available online: http://eur-lex.europa.eu/legal-content/EN/TXT/PDF/?uri=CELEX:52013DC0229&from=EN

European Commission (2014) The European Union explained -food safety. 16 p. Available online: https://europa.eu/european-union/topics/food-safety_en

European Commission (2016) Summary of the 27 Multiannual national aquaculture plans. 12 p. Available online: https://ec.europa.eu/fisheries/sites/fisheries/files/docs/body/27-multiannual-national-aquaculture-plans-summary_en.pdf

European Union (2012) The common agricultural policy-a story to be continued. Publicaetions Office of the European Union. 23 p. Available online: http://ec.europa.eu/agriculture/50-years-of-cap/files/history/history_book_lr_en.pdf

European Union (2014) Living well, within the limits of our planet, 7th EAP-The new general Union Environment Action Programme to 2020. 4 p. Available online: http://ec.europa.eu/environment/pubs/pdf/factsheets/7eap/en.pdf

European Union (2016) Commission staff working document on the application of the Water Framework Directive (WFD) and the Marine Strategy Framework Directive (MSFD) in relation to aquaculture. 36 p. Available online: http://ec.europa.eu/environment/marine/pdf/SWD_2016_178.pdf

Goddek S, Delaide B, Mankasingh U, Ragnarsdottir KV, Jijakli H, Thorarinsdottir R (2015) Challenges of sustainable and commercial aquaponics. Sustainability 7:4199–4224. https://doi.org/10.3390/su7044199

Joly A, Junge R, Bardocz T (2015) Aquaponics business in Europe: some legal obstacles and solutions. Ecocycles 1(2):3–5. https://doi.org/10.19040/ecocycles.v1i2.30

König B, Janker J, Reinhardt T, Villarroel M, Junge R (2018) Analysis of aquaponics as an emerging technological innovation system. J Clean Prod 180:232–243

Massot A (2017) The Common Agricultural Policy (CAP) and the treaty. Fact sheets on the European Union. European Parliament. Available online: http://www.europarl.europa.eu/atyourservice/en/displayFtu.html?ftuId=FTU_3.2.1.html

Paetsch U (2013) Jahresfischereitag und Jahreshauptversammlung des Landesverbandes der Binnenfischer MV e.V. – Güstrow, 25. Februar 2013: Bericht des Präsidiums für das Jahr 2012, Fischerei & Fischmarkt in MV 2/2013, 6–9

Ragonnaud G (2017) Second pillar of the CAP: rural development policy. Fact sheets on the European Union. European parliament. 4 p. Available online: http://www.europarl.europa.eu/atyourservice/en/displayFtu.html?ftuId=FTU_3.2.6.html

Schulz K, Weith T, Bokelmann W, Petzke N (2013) Urbane Landwirtschaft und "Green Production" als Teil eines nachhaltigen Landmanagements. Discussion paper 6. Müncheberg

Schendel F (2016) Abwasserabgabe – wann kommt eine Reform? Natur und Recht:166–171

Sheil S (2013) Strategic guidelines for EU aquaculture. Library briefing. 6 p. Available online: http://www.europarl.europa.eu/eplibrary/Strategic-guidelines-for-aquaculture-in-the-EU.pdf. Accessed on 05 Oct 2017

Somerville C, Cohen M, Puntanella E, Stankus A, Lovatelli A (2014) Small-scale aquaponic food production. FAO Fisheries and Aquaculture Technical paper 589. 288 p. Available online: http://www.fao.org/3/a-i4021e.pdf

Smith R, Pryor R (2015) Deliverable 5.1 – Mapping Exercise Report with regard to current Legislation & Regulation: Europe and Africa & China, 2015

Sodano V, Hingley M, Lindgreen A (2008) The usefulness of social capital in assessing the welfare effects of private and third-party certification standards: trust and networks. Br Food J 110 (4/5):493–513

Stamer A (2009) Betäubungs- & Schlachtmethoden für Speisefische, FiBL, Available online: http://orgprints.org/16511/1/stamer-2009-literaturstudie_fischschlachtung-FiBL_Bericht.pdf

Windstoßer C (2011) Rechtliche Voraussetzungen für die Errichtung von Kreislaufanlagen, Präsentation im Rahmen des Seminars "Fischproduktion in Kreislaufanlagen – Prinzipien, Wirtschaftlichkeit, Zukunft" FFS Baden-Württemberg Available online: http://www.landwirtschaft-bw.info/pb/site/lel/get/documents/MLR.LEL/PB5Documents/lazbw_ffs/Windsto%C3%9Fer_Rechtliche%20Voraussetzung.pdf

Yanong RPE, Erlacher-Reid C (2012) Biosecurity in aquaculture, Part 1: An overview. SRAC Publication No. 4707

2

Food, Sustainability, and Science Literacy in One Package? Opportunities and Challenges in using Aquaponics Among Young People at School, a Danish Perspective

Bent Egberg Mikkelsen and Collins Momanyi Bosire

Abstract The call for sustainable food production and consumption has led to an increased interest and new policy measures to support the circular economy and climate-smart farming practices. The merits of aquaponics and closed-loop nutrient cycling systems are increasingly being examined in terms of sustainable productivity in various settings including urban environments. Aquaponics also has the potential to be applied as a learning tool for people of all ages but especially for young people at school. This chapter studies the potential of aquaponics to teach food and science literacy and the use of the technology as an educational tool in primary school. The chapter draws on data from the Growing Blue & Green (GBG) program carried out in cooperation among Aalborg University, Copenhagen, municipal schools and their teachers and a private aquaponic enterprise. The chapter draws on three empirical studies including an exploratory study on the educational opportunities at school, a feasibility study carried out among teachers, as well as the educational Growing Blue & Green (eGBG) study, in which a digital-based regulation component was added. The conclusion is that low-cost versions of aquaponics have considerable potential for supportive learning in elementary school. Preliminary findings furthermore suggest that fitting the setup with easy-to-install intelligent sensors and devices offers the opportunity to provide learning about food, sustainability, and a basic understanding of the control and management of biological systems in one package.

Keywords Aquaponics · Sustainability education · Science literacy · Food literacy · Education · Primary school

---

B. E. Mikkelsen (✉)
Integrated Food Studies, Aalborg University Copenhagen, Copenhagen, Denmark
e-mail: bemi@learning.aau.dk

C. M. Bosire
Department of Learning & Philosophy, Aalborg University, Copenhagen, Denmark

Introduction

Sustainable food production and consumption are important societal challenges. Growing populations, scarcity of arable land, and urbanization are all important factors in this area, leading to a growing interest in new sustainable food production technologies, which are not necessarily confined to maritime or rural settings. Aquaponics is one of these technologies that has received increased attention, in particular, since it can easily be applied also in urban environments. These new technologies, including aquaponics, also offer new opportunities as learning tools for people of all ages, but it is particularly appealing for young people at school. This chapter reports on the findings from the educational Growing Blue & Green (GBG) program that has been developed and tested in educational settings in the Greater Copenhagen area. Additional studies also suggest that there appears to be potential for using aquaponics as a key way of learning about sustainable food production in a wide spectrum of academic disciplines at school since it can readily be integrated into the existing educational curriculum. A few studies have examined the application of aquaponics in an educational context. Graber et al. (2014) studied the potential of aquaponics as a food production method for urban areas teaching seventh-grade pupils sustainability issues in science classes. The idea behind the concept was to introduce and train students on "systems thinking" by combining fish and plant growing. Junge et al. (2014) showed that the students' ability to think in a systematic way improved significantly as a result. The study also suggested that building on social learning in groups the students developed greater teamwork skills. However, apart from these examples, aquaponics literature is relatively limited and most of the available articles have focused on the technological aspects of the systems. This chapter attempts to fill this knowledge gap by exploring the opportunities for integrating aquaponic technology into school learning and to uncover some of the constraints as well as the opportunities.

The chapter draws on three empirical cases, where aquaponics has been applied in primary school settings in the Greater Copenhagen area. This includes an exploratory study on the educational opportunities at school (Bosire et al. 2016), a feasibility study carried out among teachers (Bosire and Sikora 2017), as well as the insights from the first part of an educational Growing Blue & Green study (Toth and Mikkelsen 2018).

The aim of this chapter is thus to introduce and discuss educational aquaponic interventions in which aquaponics is used in an elementary school in Denmark and to discuss how theory can be taken into practice. The chapter discusses the potential of aquaponics to be able to contribute to the fostering of a deeper learning about urban, sustainable food production among young school children in school settings, as well as the potentials for creating digital literacy through the addition of a self-regulating digital sensing and maintenance tool, i.e., the eGBG tool.

Conceptual Foundation

Supporting Sustainable Development (SD) of the food system through educational efforts can be expected to be a good investment, as school children are the future policy makers and producers.

According to Shephard (2008), educators and particularly higher educators have traditionally focused on the cognitive domain of learning with no much emphasis being put on primary education. We hold the view that using appropriate learning tools at primary school level can be an essential pillar to bringing about long-term positive change in societies. These can be realized though alternative learning and teaching approaches, different from traditional deductive approaches such as "learning by doing" and "experiential learning" pioneered by Dewey (1997) in his work experience and education. In our research work, we present a type of extracurricular dimensional perspective, where we add to pupils' learning outcomes by tapping into the affective domain, which focuses on interests, attitudes, appreciations, values, changing behaviors, and emotional sets or biases (Shephard et al. 2015). Practical aquaponics promises to deliver a hands-on problem-based inductive learning tool for education.

Study cases all build on the idea of Service Learning (SL), where students use academic knowledge to address community needs, and the knowledge triangle (education, research, and innovation), which are part of the teaching at the Integrated Food Studies (IFS) program at Aalborg University (Mikkelsen and Justesen 2015). The IFS also uses Problem-Based Learning (PBL), where learning is approached with open-ended problems with no absolute right answer, as well SL approaches. SL is a pedagogical approach that is rooted in PBL as well as in the experiential learning approach (McKay-Nesbitt et al. 2012). Using the SL approach, students are expected to become involved in projects based on the needs, wishes, and demands of local communities. The recent interest in reforming educational practices and strategies makes the use of aquaponics an important component in the educational context timely and relevant. Additionally, the use of inductive methods such as PBL and discipline-based learning (Wood 2003: Armstrong 2008) as well as experiential learning (Beard 2010; McKay-Nesbitt et al. 2012), where everyday life problems and questions are used to inform the learning process, is spreading. These concepts are all favorable for aquaponic teaching. Furthermore, the idea of SL is compatible with the aquaponic teaching concept and the recent Danish School Reform (Danish Ministry of Education 2014) that present guidelines on how to integrate the practical and theoretical aspects of the curriculum.

While there are several aquaponic systems that can be supplied by manufacturers and/or bespoke systems designed by consultants, aquaponic technology in principle is rather simple. The basic principles can thus be well understood by students, and the systems can be designed, built, and monitored by students using a range of materials and methods, ranging from the basic to the sophisticated. Taking this

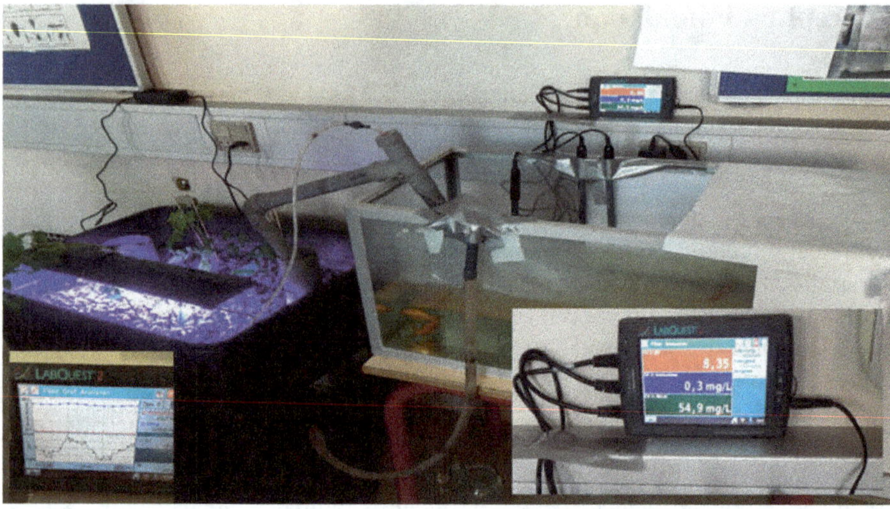

Fig. The aquaponic learning and experimental mock-up. The illustration shows the setup including aquarium fish tank and the monitoring devices that are used to measure the equilibrium of the whole system. The last part is the core of the learning goal for students. (Pictures: courtesy of Lija Gunnarsdottir)

premise, aquaponics is thus a technology that is highly suited to the knowledge triangle approach. Education can be enhanced through the creation of links between the three sides of the knowledge triangle, i.e., education, research, and innovation. Innovative thinking on how education for sustainability could be implemented using practical educational tools leads the educator toward aquaponics: a food production method which is essentially a symbiotic integration of two mature disciplines – recirculating aquaculture and hydroponics in one production system, where the live fish generate nutrients for plant production. A simple aquaponic system unit, such as the one shown in Fig., was set up at an elementary school in Copenhagen. The figure illustrates some of the basic components used with a brief detail on its working principle: a simple aquarium where water in the fish tank is kept at a constant height through appropriate design for the comfort of the fish. Through some pumping action from a sump tank situated below the grow bed, excess water containing fish waste is cycled through plant grow beds, where bacteria and other microbes are hosted.

The sump and grow bed act together as mechanical and biofilters, respectively, by removing solids and dissolved waste.

The setup in Fig. illustrates a practical educational example, focusing on sustainability since it provides a practical example of how the goals set out under Sustainable Development Goals (SDGs) in UN Agenda 2030 for SD can be addressed (UN 2015b). Goal number 2, which targets ending hunger, achieving food security and improved nutrition, and promoting sustainable agriculture, and goal number 4, which focuses on ensuring inclusive and equitable quality education and the promotion of lifelong learning opportunities for all (UN 2015b). These crucial issues can be included in the Problem-Based Learning (PBL) approach that has been developed in the GBG case. Based on a shared firm belief in having technological solutions for the problems of contemporary food systems, the GBG approach contributes to a demonstration of "ecological modernization" in food production processes. Through the development of the didactic for the eGBG themes of sustainability and food literacy, it became clear that for such a system to bring about change, there is a need for the right platform through which knowledge and skills can be exchanged among young people and their teachers in the school setting.

Other studies have shown that the lack of food and nutrition literacy among young people is of growing concern (Vidgen and Gallagos 2014; Dyg and Mikkelsen 2016). This is particularly concerning, as the conventional ways of food production and the current persistent drivers of science and technology have fueled unsustainable global exploitation of earth's resources leading to numerous challenges within the food system (FAO 2010; UNDP 2016). In addition, the increase in world population and the rapid urbanization have overloaded the food system. The United Nations predicts that world population will increase by more than 1 billion people within the next 15 years, reaching 8.5 billion in 2030. Of these, the majority (66%) is forecasted to be living in cities by the year 2050 (UN 2015a). These trends in combination with the growth in unhealthy eating habits and nutrition-related disorders have made a new approach to food nutrition and agri-literacy at school imperative.

The insights from the GBG project and the results from numerous interviews with both teachers and students showed that the successful application of aquaponic technology is dependent on the careful planning and maintenance of the system. The digital version of the GBG – the eGBG – was developed to address these challenges and to use the related opportunities in promoting digital literacy in school. The idea of the eGBG takes inspiration from the idea of self-regulation in biological systems. It is conceptually based on the idea of autopoiesis: referring to a system capable of reproducing and maintaining itself. The term first introduced in 1972 by biologists Maturana and Varela (1980) describes the self-maintaining chemistry of living cells, and ever since then, the concept has been applied in a wide array of fields such as cognition, systems theory, and sociology. In the eGBG study, illustrated by the setup and components in Fig, water quality, temperature, dissolved oxygen, CO2, pH, ammonia, and nitrite content are measured with sensors using an elec-tronic and digitalized setup, followed by appropriate automated regulation and adjustments to the required or set levels. This system used alongside a basic maintenance regime better enables children to learn Information and Communications Technology (ICT), together with Science, Technology, Engineering, and Mathematics (STEM) subjects in addition to a wider understanding of sustainable urban farming and animal welfare practices. The eGBG minimizes human error and reduces the amount of critical resources such as the physical labor and hours that would otherwise be required for the care and maintenance of a balanced aquaponic system.

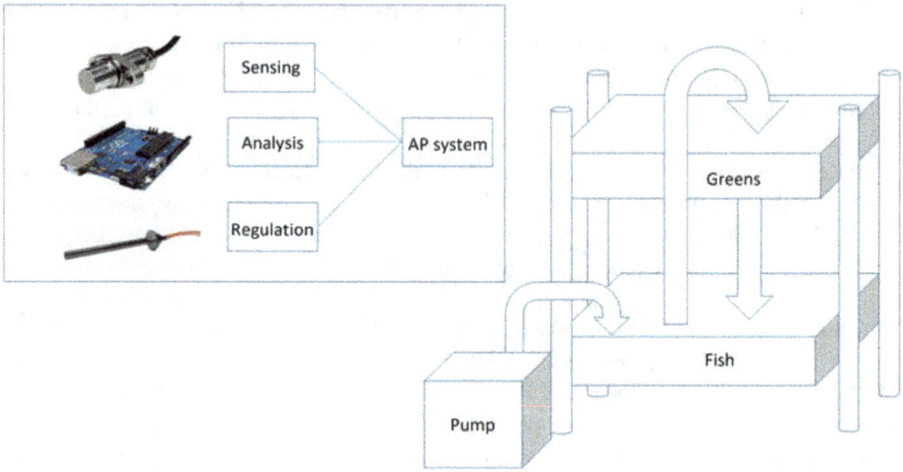

Fig. The experimental eGBG setup. The illustration shows the two parts of the system. The aquaponic system itself and the measuring devices and the minicomputer used to follow the biological condition of the eGBG system

Methods

In the context of this chapter, three data sources were used including (a) an exploratory study on the educational opportunities at school (Bosire et al. 2016), (b) a feasibility study carried out among teachers (Bosire and Sikora 2017), and (c) the eGBG study (Toth and Mikkelsen 2018).

The first study (a) was carried out as an exploration of the opportunities and challenges of using aquaponics as an educational tool. The study aimed at investigating to what extent it makes sense to use aquaponics in school teaching. Data from three (N = 3) independent qualitative interviews were collected. The informants were (1) a biology teacher engaged in natural science teaching at primary school; (2) a consultant entrepreneur, which is an aquaponic expert, too; and (3) one local aquaponic bio-farmer. The data analysis procedure was inspired by the future workshop approach (Jungk and Müllert 1987), leading to a categorization and evaluation according to the three categories of critique, fantasy, and strategy.

In the second study (b), a feasibility study was carried out at the Blågaard Public School located within the Copenhagen Municipality in cooperation with two biology teachers and a physics teacher and with approval from the school administration. The local aquaponic bio-farmer and expert was also involved. A low-cost aquaponic facility for teaching was developed using a simple do-it-yourself (DIY) food production system and off-the-shelf components. The idea for this design and construction was to illustrate that this kind of technology can readily be undertaken, and it is not only for advanced urban growing but that it also has potential to be used as a science-teaching tool in humbler settings such as a local school. Since the budget of the school is limited, the overall goal was to complete the project at low cost and to carefully fit the system within the requirements of the existing curricula.

In the third study (c), a digital component was added to an improved version of the aquaponic system and the eGBG was introduced. The eGBG is a learning program based on simple aquaponics, and it is designed to create learning insights among adolescents. The program's special focus is on teaching principles of sustainable food production in cities and at the same time to facilitate ICT learning. The program's didactics is aimed at showing how a biological system like aquaponics can be controlled, maneuvered, and self-regulated using sensors and feedback mechanisms. This is done by connecting sensors that measure temperature, pH, and nutrient balance via a digital interface such as Arduino. The eGBG developed a simple urban farming tool based on a learning package for schools where students can learn about this technology in biology classes. By studying how the sensors work, they have the ability to learn how ICT can be integrated to monitor and control a living biological system.

The eGBG educational program can be used both in an interdisciplinary course with ICT as a theme or in the subjects of biology, physics, and chemistry. The components of the eGBG are for a low-cost aquaponic system that has been developed for the school context as described previously. Some of the key elements were supplied by BioTeket, which is a company with a social and cultural remit with an emphasis on environmental technology. BioTeket offers a series of workshops and events, giving the citizens of Copenhagen an opportunity to gain experience with sustainable urban life. The assembly was done under the company's technical supervision. According to the national curricula, ICT in elementary school is taught not as a stand-alone subject, but in a transversal manner spanning several subjects. Combining smart and sensor-based control and biological system therefore seems straightforward for this requirement. Urban farming technologies require a monitoring system with a multitude of sensors since maintaining a system in balance requires continuous measurement of temperature, pH, etc. To meet this requirement, the eGBG was developed in cooperation between Aalborg University, a municipal school in Albertslund, and the enterprise BioTeket. The development process was configured as an action research study where data was collected along with the development process.

Findings and Discussion

1. From the first study, the findings showed that visions of a new way of teaching with inclusion of the modern technology could be perceived as an advantage in influencing transformational processes at school. Nevertheless, this process requires some critical, practical, and theoretical considerations for implementation of the system to make it successful and sustainable in the long term. Some of the positive issues from the users' perspectives included a wide range of application in the subjects of biology, mathematics, science, and more. Reduction

pollution and efficient resource usage; flexibility of the system setup, e.g., on rooftops; and the production of (organic-like*) twin products (fish and plant foods). Potential limitations included time constraints, lack of financial resources, as well as the need for frequent care and maintenance. (* In the EU, current legislation provides that only vegetal produce grown in soil may be considered "organic." This is not the case, e.g., in the USA, where aquaponic produce can be grown organically and legally sold as being organic.)

2. From the second study (b), the feasibility study, the experiences of the study indicated that the learning concept, the overall idea, and the didactics fit well into the educational curricula and also with the projects that the school had already planned to undertake in the field of sustainability. The experience showed that such teaching needs to be carefully planned well ahead of time. Furthermore, the idea of a knowledge triangle approach, bringing service learning, university research, a small enterprise, and the learning staff into an informal project and innovation network, is a fruitful way of organizing the undertaking. In addition, the initiative enjoys the support of the municipality that sees entrepreneurship and innovative learning approaches as important objectives.
3. The third study (c), the eGBG study, showed that the school was supportive and already had newly purchased sensors to measure pH, temperature, CO_2, and dissolved oxygen (DO). Therefore, the data could be conducted with minimal new effort for training, as the teaching staff were already well prepared to collect data digitally. The school, at the point of project startup, was already planning to measure nitrate and ammonia using the sensors, since the basic concept of the teaching was to increase knowledge, skill, and competency in relation to the nitrogen cycle. The idea of creating aquaponic technology and applying it in the teaching was readily accepted by the school since the neighboring school already had that kind of an AP system up and running.

Acknowledgments Thanks to biology teachers Mette and Else at Blågård School in Copenhagen Municipality, to Lilja Gunnarsdottir and the teachers at Herstedlund school, and to Inge Christensen from the Nature Centre in Albertslund municipality. Thanks also to Viktor Toth, a student at Integrated Food Studies, Aalborg University, for providing data from the eGBG study. Thanks also to Tomasz Sikora and Kathrine Breidahl from the Integrated Food Studies that participated in the fieldwork. Thanks also to the owner and CEO Lasse Antoni Carlsen of Bioteket, Copenhagen, for providing components and guidance in developing the GBG program.

References

Armstrong EG (2008) A hybrid model of problem-based learning. In: Boud D, Feletti G (eds) The challenge of problem-based learning. Routledge, London

Beard C (2010) The experiential learning toolkit: blending practice with concepts. Kogan Page, London

Bosire CM, Sikora TA (2017) Blue-green farms of the future: using aquaponics at Primary School to Foster sustainable development (Master thesis). Retrieved from http://projekter.aau.dk/projekter/files/259876456/Collins_Bosire_Tomasz_Sikora__Blue_Green_Farms_of_the_Future_EMT.pdf

Bosire CM, Breidahl KS, Sikora TA, Mikkelsen BE (2016) Education for sustainability and food literacy – assessing opportunities and challenges for using aquaponics among young people at Schools. In: Mikkelsen BE, Ofei KT, Tvedebrink TDO, Romani AQ, Sudzina F (eds) Proceedings from 10th international conference on culinary arts and sciences, July 5–7th 2017 Aalborg University Copenhagen – Exploring Future Foodscapes, Published by Captive Food Studies. AAU Aalborg, p 250–267

Danish Ministry of Education (2014) Improving the public school– overview of reform of standards in the Danish Public School (primary and lower secondary education) The Danish Ministry of Education. Available at http: www.uvm.dk/publicschool. ISBN (Electronic version): 978-87-603. Accessed 24 May 2016

Dewey J (1997) Experience and education. Kappa Delta Pi, Touchstone. (Original work published 1938)

Dyg PM, Mikkelsen BE (2016) Cooperation models, motivation and objectives behind farm–school collaboration: case insights from Denmark. Int J Soc Agric Food 23(1):41–62

Food and Agriculture Organization (2010) Sustainable diets and biodiversity: directions and solutions for policy, research and action. International scientific symposium, biodiversity and sustainable diets united against hunger. Rome, Italy, 3–5 November 2010

Graber A, Antenen N, Junge R (2014) The multifunctional aquaponic system at ZHAW used as research and training lab. Available at http://pd.zhaw.ch/publikation/upload/207534.pdf. Accessed 22 May 2016

Junge R, Wilhelm S, Hofstetter U (2014) Aquaponic in classrooms as a tool to promote system thinking. Available at http://www.adam-europe.eu/prj/10804/prj/Paper_VIVUSConf_Junge_et_al.pdf. Accessed 23 May 2016

Jungk R, Müllert N (1987) Future workshops: how to create desirable futures. Institute for Social Inventions, London

Maturana H, Varela F (1980) Realization of the living. In: Boston Studies in the Philosophy of Science, vol 42. Reidel, Dordrecht

McKay-Nesbitt J, DeMoranville CW, McNally D (2012) A strategy for advancing social marketing: social marketing projects in introductory marketing courses. J Soc Mark 2(1):52–69. https://doi.org/10.1108/20426761211203256

Mikkelsen BE, Justesen J (2015) Multi disciplinarily in higher educations for the captive foodscape profession: case of Denmark. In: Proceedings of the international conference on culinary arts and eating in society. Montclair State University, Montclair, pp 262–263

Shephard K (2008) Higher education for sustainability: seeking affective learning outcomes. Int J Sustain High Educ 9(1):87–98

Shephard K, John H, Brent L, Miranda M, Sheila S, Liz S, Mick S, Mary F, Tim J, Lynley D (2015) Seeking learning outcomes appropriate for 'education for sustainable development' and for higher education. Assess Eval High Educ 40(6):855–866

Toth V, Mikkelsen BE (2018) Growing blue & green – exploring the potentials of teaching sustainable urban food the smart way. Protocol for extended master thesis at Integrated Food Studies, Aalborg University

United Nations (2015a) World population prospects, the 2015 revision vol. 2: Demographic Profiles. Available at http://esa.un.org/unpd/wpp/Publications/Files/WPP2015_Volume-II-Demographic-Profiles.pdf. Accessed 8 Mar 2016

United Nations (2015b) Transforming our world: the 2030 agenda for sustainable development. Available at https://sustainabledevelopment.un.org/post2015/transformingourworld. Accessed 3 Mar 2017

United Nations Development Programme (2016) Overview: human development report 2016 human development for everyone (online). Available at http://hdr.undp.org/sites/default/files/HDR2016_EN_Overview_Web.pdf. Accessed 24 Mar 2017

Vidgen HA, Gallegos D (2014) Defining food literacy and its components. Appetite 76:50–59

Wood DF (2003) Problem based learning. BMJ 326(7384):328–330. https://doi.org/10.1136/bmj.326.7384.328

3

Insight into Risks in Aquatic Animal Health in Aquaponics

Hijran Yavuzcan Yildiz, Vladimir Radosavljevic, Giuliana Parisi, and Aleksandar Cvetkovikj

Abstract Increased public interest in aquaponics necessitates a greater need to monitor fish health to minimize risk of infectious and non-infectious disease outbreaks which result from problematic biosecurity. Fish losses due to health and disease, as well as reporting of poor management practices and quality in produce, which could in a worst-case scenario affect human health, can lead to serious economic and reputational vulnerability for the aquaponics industry. The complexity of aquaponic systems prevents using many antimicrobial/antiparasitic agents or disinfectants to eradicate diseases or parasites. In this chapter, we provide an overview of potential hazards in terms of risks related to aquatic animal health and describe preventive approaches specific to aquaponic systems.

Keywords Aquaponics · Fish health · Biosecurity · Treatment strategies

H. Yavuzcan Yildiz (✉)
Department of Fisheries and Aquaculture, Ankara University, Ankara, Turkey
e-mail: yavuzcan@ankara.edu.tr

V. Radosavljevic
Department of Fish Diseases, National Reference Laboratory for Fish Diseases, Institute of Veterinary Medicine of Serbia, Belgrade, Serbia
e-mail: vladimiradosavljevic@yahoo.co.uk

G. Parisi
Department of Agriculture, Food, Environment and Forestry (DAGRI), Animal Science Section, University of Florence, Florence, Italy
e-mail: giuliana.parisi@unifi.it

A. Cvetkovikj
Department of Parasitology and Parasitic diseases, Faculty of Veterinary Medicine, Ss. Cyril and Methodius University in Skopje, Skopje, Republic of Macedonia

Department of Fisheries, Institute of Animal Science, Ss. Cyril and Methodius University in Skopje, Skopje, Republic of Macedonia
e-mail: acvetkovic@fvm.ukim.edu.mk; acvetkovikj@gmail.com

Introduction

The European Food Safety Authority reported a variety of drivers and potential issues associated with new trends in food production, and aquaponics was identified as a new food production process/practice (Afonso et al. 2017). As a new food production process, aquaponics can be defined as 'the combination of animal aquaculture and plant culture, through a microbial link and in a symbiotic relationship'. In aquaponics, the basic approach is to get benefit from the complementary functions of the organisms and nutrient recovery. The aquaculture part of the system applies principles that are similar to recirculating aquaculture systems (RAS). Aquaponics has gained momentum due to its superior features compared to traditional production systems. Thus, aquaponics seems capable of maintaining ecosystems and strengthening capacity for adaptation to climate change, extreme weather, drought, flooding and other disasters. These attributes are within reach, but as in other agri-/aquacultural production, aquaponics is not free of risks. Given the complexity of aquaponics as an environment for co-production of aquatic animals with plants, the hazards and risks may be more complicated.

The focus in this chapter is on categories of risk (i.e. animal health/disease) rather than specific risks (e.g. flectobacillosis disease). In traditional aquaculture, some of the more common types of production risks are diseases resulting from pathogens, unsuitable water quality and system failure. Snieszko (1974) reported that infectious diseases of fish occur when susceptible fish are exposed to virulent pathogens under certain environmental conditions. Thus, the interaction of pathogens, water quality and fish resistance is linked to occurrence of disease. Previous research using risk methods has studied the routes of introduction of aquatic animal pathogens in order to secure safe trade (e.g. import risk analyses) and support biosecurity (Peeler and Taylor 2011). Considering the similarity of aquaponics to RAS, it is expected that the health problems of aquatic animals in aquaponics may be identical to aquatic animals in RAS. Specifically, fluctuations in water quality may increase susceptibility of fish to pathogens (i.e. disease-causing organisms such as virus, bacteria, parasite, fungi) in RAS and cause disease outbreaks. Microorganisms in closed systems such as RAS or aquaponics are of significance in terms of maintaining fish health. Thus, Xue et al. (2017) reported the potential correlation between fish diseases and environmental bacterial populations in RAS. High pathogen density and limited medication possibilities make the system prone to disease problems. Disease or impaired health can cause catastrophic losses with decreased survival or poor feed conversion ratios. Regardless of which potential risk becomes problematic, each has the same impact: an overall decline in the production of a marketable quality product that then results in financial loss (McIntosh 2008). Diseases can be prevented only when the risks are recognized and managed before disease occurs (Nowak 2004). The severity of risks differs and will likely change depending on when each is encountered during the production cycle.

Aquaponics and Risk: A Development Perspective for Fish Health

Fish pathogens are prevalent in the aquatic environment, and fish are generally able to resist them unless overloaded by the allostatic load (Yavuzcan Yıldız and Seçer 2017). Allostasis refers to the 'stability through change' proposed by Sterling and Eyer (1988). Put simply this is the effort of fish to maintain homeostasis through changes in physiology. Allostatic load of fish in aquaponics may be a challenging factor as aquaponics is a complex system mainly in terms of the water quality and the microbial community in the system. Hence, the diseases of fish are generally species- and system-specific. Specific aquaponic diseases have not been described yet. From aquaculture, it is known that fish diseases are difficult to detect and are usually the end result of the interaction between various factors involving the environment, nutritional status of the fish, the immune robustness of the fish, existence of an infectious agent and/or poor husbandry and management practices. In order to sustain aquaponic systems, an aquatic health management approach needs to be developed considering the species cultured, the complexity of environments in aquaponics and the type of the aquaponic system management. Profitability in aquaponic production can be affected by even small percentage decreases in production, as seen in aquaculture (Subasinghe 2005).

Aquaponics is a sustainable, innovative approach for future food production systems, but this integrated system for production currently shows difficulties in moving from the experimental stage or small-scale modules to large-scale production. It could be hypothesized that the lack of economic success of this highly sustainable production system is due to major bottlenecks not scientifically addressed yet. Without a doubt, the cost-effectiveness and technical capabilities of aquaponic systems need further research to realize a scaling up of production (Junge et al. 2017). Research activity and innovations applied since the 1980s have transformed aquaponic technology into a viable system of food production, and although small-scale plants and research-structured plants are already viable, commercial-scale aquaponics are not often economically viable. The claimed advantages attributed and recognized for aquaponic systems are the following: significant reduction in the usage of water (compared to traditional soil methods of growing plants), bigger and healthier vegetables than when grown in soil, production of plants does not require artificial fertilizer and aquaponic products are free of antibiotics, pesticides and herbicides.

Risk Analysis Overview

Risk is defined as 'uncertainty about and severity of the consequences of an activity' (Aven 2016), and the risk picture reflects (i) probabilities/frequencies of hazards/threats, (ii) expected losses given the occurrence of such a hazard/threat and (iii)

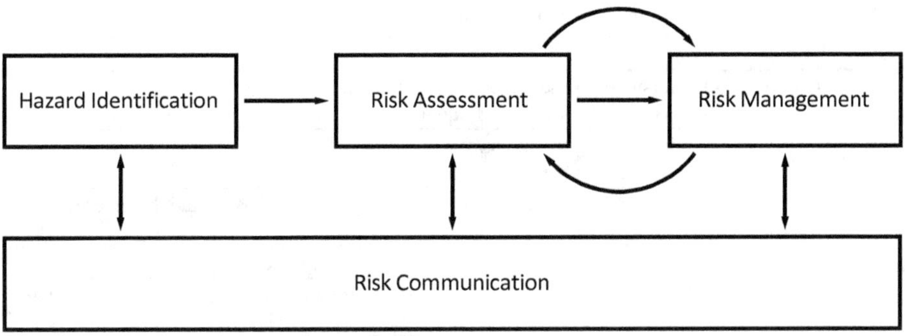

Fig. Risk analysis (OIE 2017)

factors that could create large deviations between expected outcomes and the actual outcomes (uncertainties, vulnerabilities). Risk analysis offers tools to judge risk and assist in decision-making (Ahl et al. 1993; MacDiarmid 1997). Risk analysis is based on systematic use of the available information for decision-making, using the components of hazard identification, risk assessment, risk management and risk communication as indicated by World Organisation of Animal Health (OIE) (Fig.). This framework is commonly used for pathogen risk analysis (Peeler et al. 2007).

Risk analysis in food production, including aquaponics, can be applied to many cases, such as food security, invasive species, production profitability, trade and investment, and for consumer preference for safe, high-quality products (Bondad-Reantaso et al. 2005; Copp et al. 2016). The benefits of applying risk analysis in aquaculture became more clearly linked to this sector's sustainability, profitability and efficiency, and this approach can also be effective for the aquaponics sector. Therefore, disease introduction and potential transmission of pathogens can be evaluated in the context of risk to aquatic animal health (Peeler et al. 2007). Various international agreements, conventions and protocols cover human, animal and plant health, aquaculture, wild fisheries and the general environment in the field of risk. The most comprehensive and broad agreements and protocols are the World Trade Organization's (WTO) Sanitary and Phytosanitary Agreement, United Nations Environmental Program's (UNEP) Convention on Biological Diversity and the supplementary agreement Cartagena Protocol on Biosafety and the Codex Alimentarius (Mackenzie et al. 2003; Rivera-Torres 2003).

A key challenge regarding the field of risk relates to our depth of knowledge. Risk decisions are related situations characterized by large uncertainties (Aven 2016). Specifically, animal health risk analysis depends on knowledge gained from studies of epidemiology and statistics. Oidtmann et al. (2013) point out that the main constraint in developing risk-based surveillance (RBS) designs in the aquatic context is the lack of published data to advance the design of RBS. Thus, to increase robust

Table Composite research needs for aquatic animal health in aquaponics

Research area	Research need
Basic research	Understanding the aquatic animal health and welfare concept in aquaponics in terms of the species of aquatic organisms and the system used
	Understanding the stress/stressor concept for aquatic organisms in aquaponics by the species and the system used
	Understanding the allostatic load for aquatic organisms and the emergence of diseases
	Understanding the welfare concept in aquaponics
	Characterizing the critical water quality parameters against aquatic animal health
	Understanding the sensitivity of aquatic organisms to the aquaponic environment
	Revealing the microbial profile for the different systems of the aquaponics
Health indicators	Developing and validating health indicators for aquatic animals raised in the aquaponic systems
Database development	Field data on the health/disease of aquatic animals in the aquaponics
	Field data on the microbial profile including pathogens

knowledge of risks in aquaponics, studies that both increase scientific data and reduce specific weaknesses and uncertain fields in aquaponics operations are needed. Some research areas that require more data for risk analysis in aquaponic systems are presented below (Table).

In terms of risk analysis for aquatic animal diseases or health in aquaponic systems, the OIE Aquatic Animal Health Code (the Aquatic Code) can be considered because the Aquatic Code sets out standards for the improvement of aquatic animal health and welfare of farmed fish worldwide and for safe international trade of aquatic animals and their products. This Code also includes use of antimicrobial agents in aquatic animals (OIE 2017).

Hazard Identification

In risk analysis, a hazard is generally specified by describing what might go wrong and how this might happen (Ahl et al. 1993). A hazard refers not only to the magnitude of an adverse effect but also to the likelihood of the adverse effect occurring (Müller-Graf et al. 2012). Hazard identification is important for revealing the factors that may favour the establishment of a disease and/or potential pathogen threat, or otherwise detrimental for fish welfare. Biological pathogens are recognised as hazard in aquaculture by Bondad-Reantaso et al. (2008). A broad range of factors can be taken into consideration as long as they are associated with disease occurrence, i.e. they are hazards.

Table List of potential hazards for aquatic animal health in aquaponics

	Hazard identification	Hazard specification
Abiotic	pH	Too high/too low/rapid change
	Water temperature	Too high/too low/rapid change
	Suspended solids	Too high
	Dissolved oxygen content	Too low
	Carbon dioxide content	Too high
	Ammonia content	Too high, pH dependent
	Nitrite content	Too high
	Nitrate content	Extremely high
	Metal content	Too high, pH dependent
Biotic	Stocking density	Too high/too low
	Biofouling	
Feeding	Nutrients by the fish species	Surplus/deficiency
	Feeding frequency	Inadequate/improper feeding
	Dietary toxins	
	Feed additives	Unsuitable growth promoters
Management	Aquaponic system design	Poor system design
	Fish species	Unsuitable for aquaponics
	Operational issues (water circulation, biofilter, mechanical)	
	Chemotherapeutants use	Threat for the microbial balance
	Staff hygiene	
	Biosecurity	
Welfare	Stressors	Too high
	Allostatic load	High
	Rearing conditions	Suboptimal
Diseases	Nutritional diseases	
	Environmental diseases	
	Infectious diseases	

The sustainability of aquaponics is linked with a variety of factors, including system design, fish feed and faeces features, fish welfare and elimination of pathogens from the system (Palm et al. 2014a, b). Goddek (2016) reported that aquaponic systems are characterized by a wide range of microflora as fish and biofiltration exist in the same water mass. Since a great variety of microflora exists in aquaponic practices, the occurrence of pathogens and risks for human health should also be considered in order to guarantee food safety. In terms of sustainability of aquaponic systems, pathogen elimination to prevent losses due to diseases may be a challenging factor when aquatic animal production is intensified.

The use of chemotherapeutants in aquaculture to fight pathogens presents a number of potential hazards and risks to production systems, the environment and human health (Bondad-Reantaso and Subasinghe 2008) (Table).

To eliminate hazards, the fish rearing and plant cultivation phases should be considered separately. The biggest risks in fish rearing are related to water quality, fish density, feeding quality and quantity and disease (Yavuzcan Yildiz et al. 2017). Depending on the species of fish reared, the level of risk can increase if the species is not appropriate for the conditions of the particular system. For example, potassium is often supplemented in aquaponic systems to promote plant growth, but results in reduced performance in hybrid striped bass. Normally, freshwater and high-density culture-tolerant species are utilized in aquaponics. The most common species of fish in commercial systems are tilapia and ornamental fish. Channel catfish, largemouth bass, crappies, rainbow trout, pacu, common carp, koi carp, goldfish, Asian sea bass (or barramundi) and Murray cod are among the species that have been trialled (Rakocy et al. 2006). Tilapia, a warm-water species, highly tolerant of fluctuating water parameters (pH, temperature, oxygen and dissolved solids), is the species largely reared in most commercial aquaponic systems in North America and elsewhere. The results of a recent online survey, based on answers from 257 respondents, showed that tilapia is reared in 69% of aquaponic plants (Love et al. 2015). Tilapia presents an economic interest in some markets but not in others. In the same survey (Love et al. 2015), other species utilized were ornamental fish (43%), catfish (25%), other aquatic animals (18%), perch (16%), bluegill (15%), trout (10%) and bass (7%). One of the major weaknesses in aquaponic systems is the management of water quality to meet the requirements of the tank-reared fish, while cultivated crops are treated as the second step of the process. Fish require water with appropriate parameters for oxygen, carbon dioxide, ammonia, nitrate, nitrite, pH, chlorine and others. A high level of suspended solids can affect the health status of fish (Yavuzcan Yildiz et al. 2017), provoking damages to gill structure, such as the epithelium lifting, hyperplasia in the pillar system and reduction of epithelial volume (Au et al. 2004). Fish stocking density and feeding (feeding rate and volume, feed composition and characteristics) affect the digestion processes and metabolic activities of fish and, accordingly, the catabolites, total dissolved solids (TDS) and waste by-products (faeces and uneaten feed) in the rearing water. The basic principle on which the aquaponic system is based is the utilization of catabolites in water for plant growth. Aquaponic systems require 16 essential nutrients and all these macro- and micronutrients must be balanced for optimal plant growth. An excess of one nutrient can negatively affect the bioavailability of others (Rakocy et al. 2006). Therefore, the continuous monitoring of water parameters is essential to maintain water quality appropriate for fish and crop growth and to maximize the benefits of the process. Reduced water exchange and low crop growth rate can create toxic nutrient concentrations in water for fish and crops. On the other hand, the addition of some micronutrients (Fe^{+2}, Mn^{+2}, Cu^{+2}, B^{+3} and Mo^{+6}), normally scarce in water where fish are reared, is essential to adequately sustain crop production. In comparison to hydroponic culture, crops in aquaponic systems require lower levels of total dissolved solid (TDS, 200–400 ppm) or EC (0.3–0.6 mmhos/cm) and require, like fish, a high level of dissolved oxygen in water (Rakocy et al. 2006) for root respiration.

Fish Health Management

Fish Diseases and Prevention

While fish diseases caused by bacteria, viruses, parasites or fungi can have a significant negative impact on aquaculture (Kabata 1985), the appearance of a disease in aquaponic systems can be even more devastating. Maintenance of fish health in aquaponic systems is more difficult than in RAS, and, in fact, control of fish diseases is one of the main challenges for successful aquaponics (Sirakov et al. 2016). Diseases which affect fish can be divided into two categories: infectious and non-infectious fish diseases. Infectious diseases are caused by different microbial pathogens transmitted either from the environment or from other fish. Pathogens can be transmitted between the fish (horizontal transmission) or vertically, by (externally or internally) infected eggs or infected milt. More than half of the infectious disease outbreaks in aquaculture (54.9%) are caused by bacteria, followed by viruses, parasites and fungi (McLoughlin and Graham 2007). Often, although clinical signs or lesions are not present, fish can carry pathogens in a subclinical or carrier state (Winton 2002). Fish diseases can be caused by ubiquitous bacteria, present in any water containing organic enrichment. Under certain conditions, bacteria quickly become opportunistic pathogens. The presence of low numbers of parasites on the gills or skin usually does not lead to significant health problems. The capability of a pathogen to cause clinical disease depends on the interrelationship of six major components related to fish and the environment in which they live (physiological status, host, husbandry, environment, nutrition and pathogen). If any of the components is weak, it will affect the health status of the fish (Plumb and Hanson 2011). Non-infectious diseases are usually related to environmental factors, inadequate nutrition or genetic defects (Parker 2012). Successful fish health management is accomplished through disease prevention, reduction of infectious disease incidence and reduction of disease severity when it occurs. Avoidance of contact between the susceptible fish and a pathogen should be a critical goal, in order to prevent outbreak of infectious disease.

Three main measures to achieve this goal are:

- Use of pathogen-free water supply.
- Use of certified pathogen-free stocks.
- Strict attention to sanitation (Winton 2002).

Implementation of these measures will decrease fish exposure to pathogenic agents. However, it is practically impossible to define all agents which could cause disease in the aquatic environment and to completely prevent host exposure to pathogens. Certain factors, such as overcrowding, increase fish susceptibility to infection and pathogen transmission. For that reason, many pathogens which do not cause disease in wild fish can cause disease outbreaks with high mortality rates in high-density fish production systems. To avoid this, the infection level of fish in aquaponics must be continually monitored. Maintaining biosecurity in aquaponics is important not only from an economic point of view but also for fish welfare. Appearance of any fish pathogen in constrained tank space and under high population density will inevitably pose a threat to fish health, both to the individuals that are affected by the pathogen and those still unaffected.

The goal of biosecurity is the implementation of practices and procedures which will reduce the risks of:

- Introduction of pathogens into the facility.
- Spread of pathogens throughout the facility.
- Presence of conditions which can increase susceptibility to infection and disease (Bebak-Williams et al. 2007).

The achievement of this goal involves management protocols to prevent specific pathogens from entering the production system. Quarantine is an important biosecurity component for prevention of contact with infectious agents and is used when fish are moved from one area to another. All newly acquired fish are quarantined before they are introduced into established populations. Fish under quarantine are isolated for a specific period of time before release into contact with a resident population, preferably in a separate area with dedicated equipment (Plumb and Hanson 2011). New fish remain in quarantine until shown to be disease-free. It is advisable in some cases to quarantine new fish in an isolation tank for 45 days before adding them to the main system (Somerville et al. 2014). During quarantine, fish are monitored for signs of disease and sampled for presence of infectious agents. Prophylactic treatments may be initiated during the quarantine period in order to remove initial loads of external parasites.

For disease prevention, certain measures are recommended to reduce risk factors:

- Administer commercial vaccines against various fish viral and bacterial pathogens. Most common routes of application are by injection, by immersion or via food.
- Breed strains of fish which are more resistant to certain fish pathogens. Although Evenhuis et al. (2015) report that fish strains with increased simultaneous resistance to two bacterial diseases (columnaris and bacterial cold water disease) are available, there is evidence that increased susceptibility to other pathogens may occur (Das and Sahoo 2014; Henryon et al. 2005).
- Take preventive and corrective measures to prevent stress in fish. Since multiple stressors are present in every step of aquaponic production, avoidance and management of stress through monitoring and prevention minimize its influence on fish health.

- Avoid high stocking density, which causes stress and may increase the incidence of disease even if other environmental factors are acceptable. Also, high stocking density increases the possibility of skin lesions, which are sites of various pathogen entries into the organism.
- Regularly remove contaminants from water (uneaten food, faeces and other particulate organics). Dead or dying fish should be removed promptly as they can serve as potential disease sources to the remaining stock and a breeding ground for others, as well as fouling the water when decomposing (Sitjà-Bobadilla and Oidtmann 2017).
- Disinfect all equipment used for tank cleaning and fish manipulation. After adequate disinfection, all equipment should be rinsed with clear water. Use of footbaths and hand washing with disinfecting soap at the entrance and within the buildings are recommended. These steps directly decrease the potential for the spread of pathogens (Sitjà-Bobadilla and Oidtmann 2017). Certain chemicals used as disinfectants (such as benzalkonium chloride, chloramine B and T, iodophors) are effective for disease prevention.
- Administer dietary additives and immunostimulants for improvement of health and to reduce the impacts of disease. Such diets contain various ingredients important for improvement of health and disease resistance (Anderson 1992; Tacchi et al. 2011). There exists a wide range of products and molecules, including natural plant products, immunostimulants, vitamins, microorganisms, organic acids, essential oils, prebiotics, probiotics, synbiotics, nucleotides, vitamins, etc. (Austin and Austin 2016; Koshio 2016; Martin and Król 2017).
- Segregate fish by age and species for disease prevention, since susceptibility to certain pathogens varies with age, and certain pathogens are specific to some fish species. Generally, young fish are more susceptible to pathogens than older fish (Plumb and Hanson 2011).

Maintaining the health of fish in aquaponics requires adequate health management and continuous attention. Optimal fish health is best achieved through biosecurity measures, adequate production technology and husbandry management practices which enable optimal conditions. As mentioned, avoidance through optimal rearing conditions and biosecurity procedures are the best way to avoid fish diseases. Invariably, however, a pathogen may appear in the system. The first and most important action is to identify the pathogen correctly.

Disease Diagnosis (Identification of Diseased Fish)

Early recognition of diseased fish is important in maintaining health of the aquaculture unit in the aquaponic system. Accurate diagnosis and prompt response will stop the spread of disease to other fish, thus minimizing losses.

Examination of live fish starts by observing their behaviour. Constant and careful daily observation enables early recognition of diseased fish. As a rule, fish should be observed for behavioural changes before, during and after feeding.

Healthy fish exhibit fast, energetic swimming movements and a strong appetite. They swim in normal, species-specific patterns and have intact skin without discolorations (Somerville et al. 2014). Diseased fish exhibit various behavioural changes with or without visible change in physical appearance. The most obvious indicator of deteriorating fish health is the reduction (cessation) of feeding activity, usually as a result of an environmental stress and/or an infectious/parasitic disease. The most obvious sign of disease is the presence of dead or dying animals (Parker 2012; Plumb and Hanson 2011).

Behavioural changes in diseased fish may include abnormal swimming (swimming near the surface, along the tank sides, crowding at the water inlet, whirling, twisting, darting, swimming upside down), flashing, scratching on the bottom or sides of the tank, unusually slow movement, loss of equilibrium, weakness, hanging listlessly below the surface, lying on the bottom and gasping at the water surface (sign of low oxygen level) or not reacting to external stimuli. In addition to behavioural changes, diseased fish exhibit physical signs that can be seen by the unaided eye. These gross signs can be external, internal or both and may include loss of body mass; distended abdomen or dropsy; spinal deformation; darkening or lightening of the skin; increased mucus production; discoloured areas on the body; skin erosions, ulcers or sores; fin damage; scale loss; cysts; tumours; swelling on the body or gills; haemorrhages, especially on the head and isthmus, in the eyes and at the base of fins; and bulging eyes (pop-eye, exophthalmia) or endophthalmia (sunken eyes). The internal signs are changes in the size, colour and texture of the organs or tissues, accumulation of fluids in the body cavities and presence of pathological formations such as tumours, cysts, haematomas and necrotic lesions (Noga 2010; Parker 2012; Plumb and Hanson 2011; Winton 2002).

Upon suspicion of deteriorating fish health, the first step is to check water quality (water temperature, dissolved oxygen, pH, levels of ammonia, nitrite and nitrate) and promptly respond to any deviations from the optimal range. If the majority of fish in the tank has abnormal behaviour and shows non-specific signs of disease, there is likely a change in the environmental conditions (Parker 2012; Somerville et al. 2014). Low oxygen (hypoxia) is a frequent cause of fish mortality. Fish in water with low oxygen are lethargic, congregate near the water surface, gasp for air and have brighter pigmentation. Dying fish exhibit agonal respiration, with mouth open and opercula flared. These signs are also evident in fish carcasses. High ammonia levels cause hyperexcitability with muscular spasms, cessation of feeding and death. Chronic deviation from optimal levels results in anaemia and decreased growth and disease resistance. Nitrite-poisoned fish have behavioural changes characteristic of hypoxia with pale tan or brown gills and brown blood (Noga 2010).

When only few fish show signs of disease, it is imperative to remove them immediately in order to stop and prevent the spread of the disease agent to the other fish. In the early stages of a disease outbreak, generally only a few fish will show signs and die. In the following days, there will be a gradual increase in the daily mortality rate. The diseased fish must be carefully examined in order to determine the cause. Only a few fish diseases produce pathognomonic (specific to a given disease) behavioural and physical signs. Nevertheless, careful observation will often allow the examiner to narrow down the cause to environmental conditions or disease agents. In a serious disease outbreak, a fish veterinarian/health specialist should be contacted immediately for professional diagnosis and disease management options. In order to solve the disease problem, the diagnostician will need a detailed description of the behavioural and physical signs exhibited by the diseased fish, daily records of the water quality parameters, origin of the fish, date and size of fish at stocking, feeding rate, growth rate and daily mortality (Parker 2012; Plumb and Hanson 2011; Somerville et al. 2014).

Treatment Strategies in Aquaponics

Treatment options for diseased fish in an aquaponic system are very limited. As both fish and plants share the same water loop, medications used for disease treatments can easily harm or destroy the plants, and some may get absorbed by the plants, causing withdrawal periods or even making them unusable for consumption. The medications can also have detrimental effects on the beneficial bacteria in the system. If a medicinal treatment is absolutely necessary, it must be implemented early in the course of the disease. The diseased fish is transferred into a separate (hospital, quarantine) tank isolated from the system for treatment. When returning the fish after the treatment, it is important not to transfer the used medications into the aquaponic system. All these limitations require improvements of disease management options with minimal negative effects to the fish, the plants and the system (Goddek et al. 2015, 2016; Somerville et al. 2014; Yavuzcan Yildiz et al. 2017). One of the most used and effective, old-school treatments against the most common bacterial, fungal and parasitic infections in fish is a salt (sodium chloride) bath. Salt is beneficial for the fish, but can be detrimental to the plants in the system (Rakocy 2012), and the whole treatment procedure must be performed in a separate tank. A good option is to separate the recirculating aquaculture unit from the hydroponic unit (decoupled aquaponic systems). Decoupling allows for fish disease and water treatment options that are not possible in coupled systems (Monsees et al. 2017). One recent improvement for the control of fish ectoparasites and disinfection in the aquaponic systems is the use of Wofasteril (KeslaPharmaWolfen GMBH, Bitterfeld-Wolfen, Germany), a peracetic acid-containing product that leaves no residues in the system (Sirakov et al. 2016). Alternatively, hydrogen peroxide can be used, but at a much higher concentration. While these chemicals have minimal side effects, their presence is undesirable in aquaponic systems and alternative approaches, such as biological control methods, are required (Rakocy 2012).

The biological control method (biocontrol) is based on the use of other living organisms in the system, relying on natural relationships among the species (commensalism, predation, antagonism, etc.) (Sitjà-Bobadilla and Oidtmann 2017) to control fish pathogens. At present, this method is a complementary fish health management tool with high potential, especially in aquaponic systems. The most successful implementation of biocontrol in fish culture is the use of cleaner fish against sea lice (skin parasites) in salmon farms. It is best practiced in Norwegian farms where cleaning wrasse (Labridae) are co-cultured with salmon. The wrasse remove and feed on sea lice (Skiftesvik et al. 2013). Although cleaning is less common in freshwater fish, the leopard plecos (*Glyptoperichthys gibbiceps*), cohabiting with blue tilapia (*Oreochromis aureus*), successfully keeps infection with *Ichthyophthirius multifiliis* under control by feeding on the parasite cysts (Picón-Camacho et al. 2012). This biocontrol method is becoming increasingly important in aquaculture and can be considered in aquaponic systems. Additionally, it must be noted that the cleaner fish can also harbour pathogens that can be transmitted to the main cultured species. Therefore, they must also undergo preventive and quarantine procedures before introduction into the system.

Another biocontrol method, still in the exploratory application phase in fish culture, is the use of filter-feeding and filtering organisms. By reducing the pathogen loads in the water, these organisms can lower the chances of disease emergence (Sitjà-Bobadilla and Oidtmann 2017). For example, Othman et al. (2015) demonstrated the ability of freshwater mussels (*Pilsbryoconcha exilis*) to reduce the population of *Streptococcus agalactiae* in a laboratory-scale tilapia culture system. The potential of this biocontrol method in aquaponic systems is yet to be tested, and new studies are needed to explore the possibilities not only for fish disease control but also for control of plant pathogens.

The most promising and well-documented biocontrol method is the use of beneficial microorganisms as probiotics in fish feed or in the rearing water. Their usage in aquaponic systems as promoters of fish/plant growth and health is well known, and probiotics have also demonstrated effectiveness against a range of bacterial pathogens in different fish species. For example, in rainbow trout, dietary *Carnobacterium maltaromaticum* and *C. divergens* protected from *Aeromonas salmonicida* and *Yersinia ruckeri* infections (Kim and Austin 2006) and *Aeromonas sobria* GC2 incorporated into the feed successfully prevented clinical disease caused by *Lactococcus garvieae* and *Streptococcus iniae* (Brunt and Austin 2005). Dietary *Micrococcus luteus* reduced the mortalities from *Aeromonas hydrophila* infection and enhanced the growth and health of Nile tilapia (Abd El-Rhman et al. 2009). Recent research by Sirakov et al. (2016) has made good progress in simultaneous biocontrol of parasitic fungi in both fish and plants in a closed recirculating aquaponic system. In total, over 80% of the isolates (bacteria isolated from the aquaponic system) were antagonistic to both fungi (*Saprolegnia parasitica* and *Pythium ultimum*) in the *in vitro* tests. Bacteria were not classified taxonomically, and the authors assumed that they belonged to the genus *Pseudomonas* and to a group of lactic acid bacteria. These findings, although very promising, have yet to be tested in an operational aquaponic system.

As a final alternative to chemical treatment, we suggest the use of medicinal plants with antibacterial, antiviral, antifungal and antiparasitic properties. Plant extracts have various biological characteristics with minimal risk of developing resistance in the targeted organisms (Reverter et al. 2014). Many scientific reports demonstrate the effectiveness of medicinal plants against fish pathogens. For example, Nile tilapia fed with a diet containing mistletoe (*Viscum album coloratum*) increased the survivability when challenged with *Aeromonas hydrophila* (Park and Choi 2012). Indian major carp showed a significant reduction in mortality when challenged with *Aeromonas hydrophila* and fed with diets containing prickly chaff flower (*Achyranthes aspera*) and Indian ginseng (*Withania somnifera*) (Sharma et al. 2010; Vasudeva Rao et al. 2006). Medicinal plant extracts have also proven effective against ectoparasites. In goldfish, Yi et al. (2012) demonstrated the effectiveness of *Magnolia officinalis* and *Sophora alopecuroides* extracts against *Ichthyophthirius multifiliis*, and Huang et al. (2013) showed that extracts of *Caesalpinia sappan*, *Lysimachia christinae*, *Cuscuta chinensis*, *Artemisia argyi* and *Eupatorium fortunei* have 100% anthelmintic efficacy against *Dactylogyrus intermedius*. The use of medicinal plants in aquaponics is promising, but yet more research is needed to find the appropriate treatment strategy without undesirable effects. As referred by Junge et al. (2017), even though research on aquaponics has largely developed in recent years, the number of research papers published on the topic is still dramatically low compared to papers published related to aquaculture or hydroponics. Aquaponics, still considered an emerging technology, is however now characterized by having great potential for food production for the world's population that, according to the results of the UN World Population Prospects (UN 2017), numbered nearly 7.6 billion in mid-2017 and, based on the projections, it is expected to increase to 1 billion within 12 years, reaching about 8.6 billion in 2030. Nevertheless, considering the potential risks to the sustainability of aquaponics due to fish diseases, development of good ideas, and novel methods and approaches for pathogen control will be our major challenge for the future. There is a pressing need to initiate new knowledge to provide a better basis for management of fish and plant health, and to continue to develop operation and infrastructure systems for the aquaponic industry. The causes of fish losses in aquaponic systems, system-specific diseases and the interaction and alteration of microbial community, along with pathogens, are priority areas for study.

References

Abd El-Rhman AM, Khattab YAE, Shalaby AME (2009) *Micrococcus luteus* and *Pseudomonas* species as probiotics for promoting the growth performance and health of Nile tilapia, *Oreochromis niloticus*. Fish Shellfish Immunol 27:175–180. https://doi.org/10.1016/j.fsi.2009.03.020

Afonso A, Matas RG, Maggiore A, Merten C, Robinson T (2017) EFSA's activities on emerging risks in 2016. EFSA Support Publ 14:1–59. https://doi.org/10.2903/sp.efsa.2017.EN-1336

Ahl AS, Acree JA, Gipson PS, McDowell RM, Miller L, McElvaine MD (1993) Standardization of nomenclature for animal health risk analysis. Rev Sci Tech 12:1045–1053

Anderson DP (1992) Immunostimulants, adjuvants, and vaccine carriers in fish: applications to aquaculture. Annu Rev Fish Dis 2:281–307. https://doi.org/10.1016/0959-8030(92)90067-8

Au DWT, Pollino CA, Wu RSS, Shin PKS, Lau STF, Tang JYM (2004) Chronic effects of suspended solids on gill structure, osmoregulation, growth, and triiodothyronine in juvenile green grouper *Epinephelus coioides*. Mar Ecol Prog Ser 266:255–264. https://doi.org/10.3354/meps266255

Austin B, Austin DA (2016) Bacterial fish pathogens: disease of farmed and wild fish, 6th edn. Springer, Cham. https://doi.org/10.1007/978-3-319-32674-0

Aven T (2016) Risk assessment and risk management: review of recent advances on their foundation. Eur J Oper Res 253:1–13. https://doi.org/10.1016/j.ejor.2015.12.023

Bebak-Williams J, Noble A, Bowser P, Wooster G (2007) Fish health management. In: Timmons MB, Ebeling JM (eds) Recirculation aquaculture. Cayuga Aqua Ventures, Ithaca, NY, and Northeastern Regional Aquaculture Center. Publication No. 01-007, p 619–664

Bondad-Reantaso MG, Subasinghe RP (2008) Meeting the future demand for aquatic food through aquaculture: the role of aquatic animal health. In: Fisheries for global welfare and environment, 5th World Fisheries Congress 2008, pp 197–207

Bondad-Reantaso MG, Subasinghe RP, Arthur JR, Ogawa K, Chinabut S, Adlard R, Tan Z, Shariff M (2005) Disease and health management in Asian aquaculture. Vet Parasitol 132:249–272. https://doi.org/10.1016/j.vetpar.2005.07.005

Bondad-Reantaso MG, Arthur JR, Subasinghe RP (2008) Understanding and applying risk analysis in aquaculture, FAO Fisheries and Aquaculture Technical Paper No. 519

Brunt J, Austin B (2005) Use of a probiotic to control lactococcosis and streptococcosis in rainbow trout, *Oncorhynchus mykiss* (Walbaum). J Fish Dis 28:693–701. https://doi.org/10.1111/j.1365-2761.2005.00672.x

Copp GH, Russell IC, Peeler EJ, Gherardi F, Tricarico E, Macleod A, Cowx IG, Nunn AD, Occhipinti-Ambrogi A, Savini D, Mumford J, Britton JR (2016) European non-native species in aquaculture risk analysis scheme – a summary of assessment protocols and decision support tools for use of alien species in aquaculture. Fish Manag Ecol 23:1–11. https://doi.org/10.1111/fme.12074

Das S, Sahoo PK (2014) Markers for selection of disease resistance in fish: a review. Aquac Int 22:1793–1812. https://doi.org/10.1007/s10499-014-9783-5

Evenhuis JP, Leeds TD, Marancik DP, LaPatra SE, Wiens GD (2015) Rainbow trout (*Oncorhynchus mykiss*) resistance to columnaris disease is heritable and favorably correlated with bacterial cold water disease resistance. J Anim Sci 93:1546–1554. https://doi.org/10.2527/jas.2014-8566

Goddek S (2016) Three-loop aquaponics systems: chances and challenges. In: Proceedings of the international conference on Aquaponics Research Matters, Ljubljana, Slovenia, 22 March 2016

Goddek S, Delaide B, Mankasingh U, Ragnarsdottir K, Jijakli H, Thorarinsdottir R (2015) Challenges of sustainable and commercial aquaponics. Sustainability 7:4199–4224. https://doi.org/10.3390/su7044199

Goddek S, Espinal C, Delaide B, Jijakli M, Schmautz Z, Wuertz S, Keesman K (2016) Navigating towards decoupled aquaponic systems: a system dynamics design approach. Water 8:303. https://doi.org/10.3390/w8070303

Henryon M, Berg P, Olesen NJ, Kjær TE, Slierendrecht WJ, Jokumsen A, Lund I (2005) Selective breeding provides an approach to increase resistance of rainbow trout (*Oncorhynchus mykiss*) to the diseases, enteric redmouth disease, rainbow trout fry syndrome, and viral haemorrhagic septicaemia. Aquaculture 250:621–636. https://doi.org/10.1016/j.aquaculture.2004.12.022

Huang A-G, Yi Y-L, Ling F, Lu L, Zhang Q-Z, Wang G-X (2013) Screening of plant extracts for anthelmintic activity against *Dactylogyrus intermedius* (Monogenea) in goldfish (*Carassius auratus*). Parasitol Res 112:4065–4072. https://doi.org/10.1007/s00436-013-3597-7

Junge R, König B, Villarroel M, Komives T, Jijakli MH (2017) Strategic points in aquaponics. Water 9:1–9. https://doi.org/10.3390/w9030182

Kabata Z (1985) Parasites and diseases of fish cultured in the tropics. Taylor & Francis, London

Kim D-H, Austin B (2006) Innate immune responses in rainbow trout (*Oncorhynchus mykiss*, Walbaum) induced by probiotics. Fish Shellfish Immunol 21:513–524. https://doi.org/10.1016/j.fsi.2006.02.007

Koshio S (2016) Immunotherapies targeting fish mucosal immunity – current knowledge and future perspectives. Front Immunol 6:643. https://doi.org/10.3389/fimmu.2015.00643

Love DC, Fry JP, Li X, Hill ES, Genello L, Semmens K, Thompson RE (2015) Commercial aquaponics production and profitability: findings from an international survey. Aquaculture 435:67–74. https://doi.org/10.1016/j.aquaculture.2014.09.023

MacDiarmid SC (1997) Risk analysis, international trade, and animal health. In: Molak V (ed) Fundamentals of risk analysis and risk management. CRC Lewis Publishers, Boca Raton, pp 377–387

Mackenzie R, Burhenne-Guilmin F, La Viña AGM, Werksman JD, Ascencio A, Kinderlerer J, Kummer K, Tapper R (2003) An explanatory guide to the Cartagena protocol on biosafety. IUCN, Gland and Cambridge, UK, p 295

Martin SAM, Król E (2017) Nutrigenomics and immune function in fish: new insights from omics technologies. Dev Comp Immunol 75:86–98. https://doi.org/10.1016/j.dci.2017.02.024

McIntosh D (2008) Aquaculture risk management. NRAC Publication No. 107. 1 Feb 2018. http://www.mdsg.umd.edu/sites/default/files/files/107-Risk%20management.pdf

McLoughlin MF, Graham DA (2007) Alphavirus infections in salmonids - a review. J Fish Dis 30:511–531. https://doi.org/10.1111/j.1365-2761.2007.00848.x

Monsees H, Kloas W, Wuertz S (2017) Decoupled systems on trial: eliminating bottlenecks to improve aquaponic processes. PLoS One 12:e0183056. https://doi.org/10.1371/journal.pone.0183056

Müller-Graf C, Berthe F, Grudnik T, Peeler E, Afonso A (2012) Risk assessment in fish welfare, applications and limitations. Fish Physiol Biochem 38:231–241. https://doi.org/10.1007/s10695-011-9520-1

Noga E (2010) Fish disease: diagnosis and treatment, 2nd edn. Wiley-Blackwell, Ames. https://doi.org/10.1002/9781118786758

Nowak BF (2004) Assessment of health risks to southern bluefin tuna under current culture conditions. Bull Eur Assoc Fish Pathol 24:45–51

Oidtmann B, Peeler E, Lyngstad T, Brun E, Bang Jensen B, Stärk KDC (2013) Risk-based methods for fish and terrestrial animal disease surveillance. Prev Vet Med 112:13–26. https://doi.org/10.1016/j.prevetmed.2013.07.008

OIE (2017) Aquatic animal health code. Office International des Epizooties, Paris. 1 Feb 2018. http://www.oie.int/index.php?id=171&L=0&htmfile=preface.htm

Othman F, Islam MS, Sharifah EN, Shahrom-Harrison F, Hassan A (2015) Biological control of streptococcal infection in Nile tilapia *Oreochromis niloticus* (Linnaeus, 1758) using filter-feeding bivalve mussel *Pilsbryocon chaexilis* (Lea, 1838). J Appl Ichthyol 31:724–728. https://doi.org/10.1111/jai.12804

Palm HW, Seidemann R, Wehofsky S, Knaus U (2014a) Significant factors affecting the economic sustainability of closed aquaponic systems. Part I: system design, chemo-physical parameters and general aspects. AACL Bioflux 7:20–32

Palm HW, Bissa K, Knaus U (2014b) Significant factors affecting the economic sustainability of closed aquaponic systems. Part II: fish and plant growth. AACL Bioflux 7:162–175

Park K-H, Choi S-H (2012) The effect of mistletoe, *Viscum album coloratum*, extract on innate immune response of Nile tilapia (*Oreochromis niloticus*). Fish Shellfish Immunol 32:1016–1021. https://doi.org/10.1016/j.fsi.2012.02.023

Parker R (2012) Aquaculture science, 3rd edn. Delmar, Cengage Learning, Clifton Park

Peeler EJ, Taylor NG (2011) The application of epidemiology in aquatic animal health -opportunities and challenges. Vet Res 42:94. https://doi.org/10.1186/1297-9716-42-94

Peeler EJ, Murray AG, Thebault A, Brun E, Giovaninni A, Thrush MA (2007) The application of risk analysis in aquatic animal health management. Prev Vet Med 81:3–20. https://doi.org/10.1016/j.prevetmed.2007.04.012

Picón-Camacho SM, Leclercq E, Bron JE, Shinn AP (2012) The potential utility of the leopard pleco (*Glyptoperichthys gibbiceps*) as a biological control of the ciliate protozoan *Ichthyophthirius multifiliis*. Pest Manag Sci 68:557–563. https://doi.org/10.1002/ps.2293

Plumb JA, Hanson LA (2011) Health maintenance and principal microbial diseases of cultured fishes, 3rd edn. Wiley-Blackwell, Oxford. https://doi.org/10.1002/9780470958353

Rakocy J (2012) Aquaponics-integrating fish and plant culture. In: Tidwell J (ed) Aquaculture production systems. Wiley-Blackwell, Ames, pp 343–386. https://doi.org/10.1002/9781118250105

Rakocy JE, Masser MP, Losordo TM (2006) Recirculating aquaculture tank production systems: aquaponics- integrating fish and plant culture. SRAC Publication No. 454. 1 Feb 2018. http://www.aces.edu/dept/fisheries/aquaculture/documents/309884-SRAC454.pdf

Reverter M, Bontemps N, Lecchini D, Banaigs B, Sasal P (2014) Use of plant extracts in fish aquaculture as an alternative to chemotherapy: current status and future perspectives. Aquaculture 433:50–61. https://doi.org/10.1016/j.aquaculture.2014.05.048

Rivera-Torres O (2003) The biosafety protocol and the WTO, 26 B.C. Int Comp L Rev 263, http://lawdigitalcommons.bc.edu/iclr/vol26/iss2/7

Sharma A, Deo AD, Riteshkumar ST, Chanu TI, Das A (2010) Effect of *Withania somnifera* (L. Dunal) root as a feed additive on immunological parameters and disease resistance to *Aeromonas hydrophila* in Labeo rohita(Hamilton) fingerlings. Fish Shellfish Immunol 29:508–512. https://doi.org/10.1016/j.fsi.2010.05.005

Sirakov I, Lutz M, Graber A, Mathis A, Staykov Y, Smits T, Junge R (2016) Potential for combined biocontrol activity against fungal fish and plant pathogens by bacterial isolates from a model aquaponic system. Water 8:518. https://doi.org/10.3390/w8110518

Sitjà-Bobadilla A, Oidtmann B (2017) Integrated pathogen management strategies in fish farming. In: Jeney G (ed) Fish diseases: prevention and control strategies. Academic, London, pp 119–144. https://doi.org/10.1016/B978-0-12-804564-0.00005-3

Skiftesvik AB, Bjelland RM, Durif CMF, Johansen IS, Browman HI (2013) Delousing of Atlantic salmon (*Salmo salar*) by cultured vs. wild ballan wrasse (*Labrus bergylta*). Aquaculture 402–403:113–118. https://doi.org/10.1016/j.aquaculture.2013.03.032

Snieszko SF (1974) The effects of environmental stress on outbreaks of infectious diseases of fishes. J Fish Biol 6:197–208

Somerville C, Cohen M, Pantanella E, Stankus A, Lovatelli A (2014) Small-scale aquaponic food production. Integrated fish and plant farming. FAO Fisheries and Aquaculture, Rome

Sterling P, Eyer J (1988) Allostasis: a new paradigm to explain arousal pathology. In: Fischer S, Reason J (eds) Handbook of life stress, cognition and health. Wiley, Chichester, pp 629–639

Subasinghe RP (2005) Epidemiological approach to aquatic animal health management: opportunities and challenges for developing countries to increase aquatic production through aquaculture. Prev Vet Med 67:117–124. https://doi.org/10.1016/j.prevetmed.2004.11.004

Tacchi L, Bickerdike R, Douglas A, Secombes CJ, Martin SAM (2011) Transcriptomic responses to functional feeds in Atlantic salmon (*Salmo salar*). Fish Shellfish Immunol 31:704–715. https://doi.org/10.1016/j.fsi.2011.02.023

UN (2017) United Nations, Department of Economic and Social Affairs, Population Division, World Population Prospects 2017 - Data Booklet (ST/ESA/SER.A/401). 1 Feb 2018. https://esa.un.org/unpd/wpp/publications/Files/WPP2017_DataBooklet.pdf

Vasudeva Rao Y, Das BK, Jyotyrmayee P, Chakrabarti R (2006) Effect of *Achyranthes aspera* on the immunity and survival of *Labeo rohita* infected with *Aeromonas hydrophila*. Fish Shellfish Immunol 20:263–273. https://doi.org/10.1016/j.fsi.2005.04.006

Winton J (2002) Fish health management. In: Wedemayer G (ed) Fish hatchery management, 2nd edn. American Fisheries Society, Bethesda, pp 559–640

Xue S, Xu W, Wei J, Sun J (2017) Impact of environmental bacterial communities on fish health in marine recirculating aquaculture systems. Vet Microbiol 203:34–39. https://doi.org/10.1016/j.vetmic.2017.01.034

Yavuzcan Yıldız H, Seçer SF (2017) Stress and fish health: towards an understanding of allostatic load. In: Berillis P (ed) Trends in fisheries and aquatic animal health. Bentham Science Publishers, Sharjah, pp 133–154. https://doi.org/10.2174/97816810858071170101

Yavuzcan Yildiz H, Robaina L, Pirhonen J, Mente E, Domínguez D, Parisi G (2017) Fish welfare in aquaponic systems: its relation to water quality with an emphasis on feed and faeces-a review. Water 9:13. https://doi.org/10.3390/w9010013

Yi Y-L, Lu C, Hu X-G, Ling F, Wang G-X (2012) Antiprotozoal activity of medicinal plants against *Ichthyophthirius multifiliis* in goldfish (*Carassius auratus*). Parasitol Res 111:1771–1778. https://doi.org/10.1007/s00436-012-3022-7

4

Fish Diets in Aquaponics

Lidia Robaina, Juhani Pirhonen, Elena Mente, Javier Sánchez, and Neill Goosen

Abstract Fish and feed waste provide most of the nutrients required by the plants in aquaponics if the optimum ratio between daily fish feed inputs and the plant growing area is sustained. Thus, the fish feed needs to fulfil both the fish's and plant's nutritional requirements in an aquaponic system. A controlled fish waste production strategy where the nitrogen, phosphorus and mineral contents of fish diets are manipulated and used provides a way of influencing the rates of accumulation of nutrients, thereby reducing the need for the additional supplementation of nutrients. To optimize the performance and cost-effectiveness of aquaponic production, fish diets and feeding schedules should be designed carefully to provide nutrients at the right level and time to complement fish, bacteria and plants. To achieve this, a species-specific tailor-made aquaponic feed may be optimized to suit the aquaponic system as a whole. The optimal point would be determined based on overall system performance parameters, including economic and environmental sustainability measures. This chapter thus focuses on fish diets and feed and reviews the state of the art

L. Robaina (✉)
Aquaculture Research Group (GIA), Ecoaqua Institute, University of Las Palmas de Gran Canaria, Telde, Gran Canaria, Spain
e-mail: lidia.robaina@ulpgc.es

J. Pirhonen
Department of Biological and Environmental Science, University of Jyväskylä, Jyväskylä, Finland
e-mail: juhani.pirhonen@jyu.fi

E. Mente
Department of Ichthyology and Aquatic Environment, University of Thessaly, Volos, Greece
e-mail: emente@uth.gr

J. Sánchez
Department of Physiology, Faculty of Biology, Regional Campus of International Excellence "Campus Mare Nostrum", University of Murcia, Murcia, Spain
e-mail: javisan@um.es

N. Goosen
Department of Process Engineering, Stellenbosch University, Stellenbosch, South Africa
e-mail: njgoosen@sun.ac.za

in fish diets, ingredients and additives, as well as the nutritional/sustainable challenges that need to be considered when producing specific aquaponic feeds.

Keywords Aquaponic diets · Sustainability · Feed by-products · Nutrient flow · Nutritional requirements · Feeding times

Introduction

Aquatic food is recognized to be beneficial to human nutrition and health and will play an essential role in future sustainable healthy diets (Beveridge et al. 2013). In order to achieve this, the global aquaculture sector must contribute to increasing the quantity and quality of fish supplies between now and 2030 (Thilsted et al. 2016). This growth should be promoted not only by increasing the production and/or number of species but also by systems diversification. However, fish from aquaculture has only recently been included in the food security and nutrition (FSN) debate and the future strategies and policies, demonstrating the important role of this production to prevent malnutrition in the future (Bénét et al. 2015), as fish provide a good source of protein and unsaturated fats, as well as minerals and vitamins. It is important to note that many African nations are promoting aquaculture as the answer to some of their current and future food production challenges. Even in Europe, fish supply is currently not self-sufficient (with an unbalanced domestic supply/demand), being increasingly dependent on imports. Therefore, ensuring the successful and sustainable development of global aquaculture is an imperative agenda for the global and European economy (Kobayashi et al. 2015). Sustainability is generally required to show three key aspects: environmental acceptability, social equitability and economic viability. Aquaponic systems provide an opportunity to be sustainable, by combining both animal and plant production systems in a cost-efficient, environmentally friendly and socially beneficial ways. For Staples and Funge-Smith (2009), sustainable development is the balance between ecological well-being and human well-being, and in the case of aquaculture, an ecosystem approach has been only recently understood as a priority area for research.

Aquaculture has been the fastest growing food production sector during the last 40 years (Tveterås et al. 2012), being one of the most promising farming activities to meet near-future world food needs (Kobayashi et al. 2015). Total production statistics from aquaculture (FAO 2015) reveal an annual increment in global production of 6%, which is expected to provide up to 63% of global fish consumption by 2030 (FAO 2014), for an estimated population of nine billion people in 2050. In the case of Europe, the predicted increase is seen not only within the marine sector but also in terrestrially produced products. Some of the expected challenges to the growth of aquaculture during the coming years are the reduction in the use of antibiotics and other pathological treatments, the development of efficient aquaculture systems and equipment, together with species diversification and increased sustainability in the area of feed production and feed use. The shift from fishmeal (FM) in feed to other protein sources is also an important challenge, as well as the 'fish-in-fish-out' ratios. There is a long history, reaching back to the 1960s, of promoting the growth of the aquaculture sector towards proper sustainability including the encouragement to adapt and create new and more sustainable feed formulas, reduce feed spilling and reduce the food conversion ratio (FCR). Although aquaculture is recognized as the most efficient animal production sector, when compared with terrestrial animal production, there is still room for improvement in terms of resource efficiency, diversification of species or methods of production, and moreover a clear need for an ecosystem approach taking full advantage of the biological potential of the organisms and providing adequate consideration of environmental and societal factors (Kaushik 2017). This growth in aquaculture production will need to be supported by an increase in the expected total feed production. Approximately three million additional tons of feed will need to be produced each year to support the expected aquaculture growth by 2030. Moreover, replacing fishmeal and fish oil (FO) with plant and terrestrial substitutes is needed which requires essential research into formula feed for animal farming.

The animal and aquafeed industries are part of a global production sector, which is also the focus for future development strategies. Alltech's yearly survey (Alltech 2017) reveals that total animal feed production broke through 1 billion metric tons, with a 3.7% increase in production from 2015 despite a 7% decrease in the number of feed mills. China and the USA dominated production in 2016, accounting for 35% of the world's total feed production. The survey indicates that the top 10 producing countries have more than half of the world's feed mills (56%) and account for 60% of total feed production. This concentration in production means that many of the key ingredients traditionally used in formulations for commercial aquaculture feeds are internationally traded commodities, which subjects aquafeed production to any global market volatility. Fishmeal for example is expected to double in price by 2030, whilst fish oil is likely to increase by over 70% (Msangi et al. 2013). This illustrates the importance of reducing the amount of these ingredients in fish feed whilst increasing the interest and focus on new or alternative sources (García-Romero et al. 2014a, b; Robaina et al. 1998, 1999; Terova et al. 2013; Torrecillas et al. 2017).

Whilst new offshore platforms have been developed for aquaculture production, there is also a significant focus on marine and freshwater recirculating aquaculture systems (RAS), as these systems use less water per kg fish feed used, which increases fish production whilst reducing environmental impacts of aquaculture including reductions in water usage (Ebeling and Timmons 2012; Kingler and Naylor 2012). RAS can be integrated with plant production in aquaponic systems, which readily fit into local and regional food system models (see Chapter) that can be practised in or near large population centres (Love et al. 2015a). Water, energy and fish feed are the three largest physical inputs for aquaponic systems (Love et al. 2014, 2015b). Approximately 5% of feed is not consumed by the farmed fish, whilst the remaining 95% is ingested and digested (Khakyzadeh et al. 2015). Of this share, 30–40% is retained and converted into new biomass, whilst the

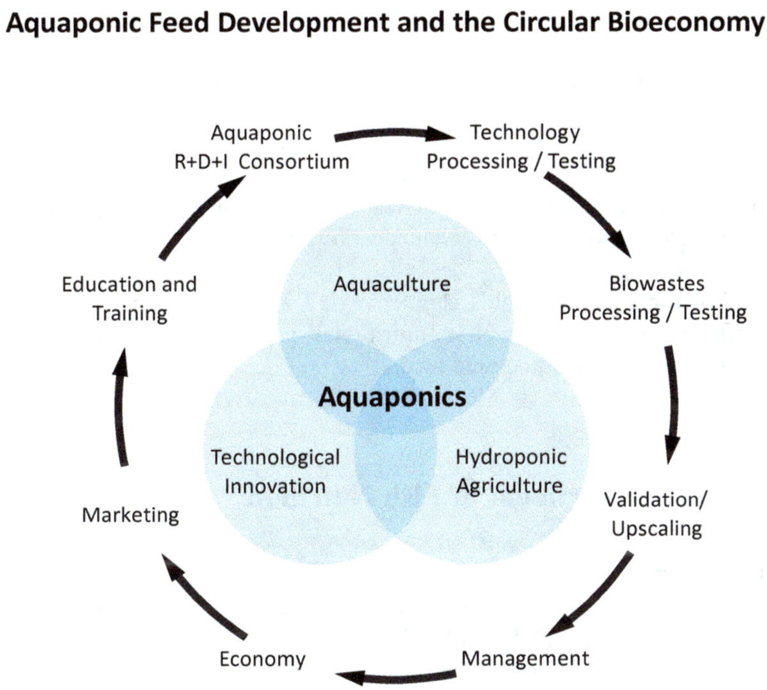

Fig. Schematic representation of a multidisciplinary approach to locally valorize bio-by-products for aquaponic diets. (Based on 'R+D+I towards aquaponic development in the ultraperipheric islands and the circular economy'; ISLANDAP project, Interreg MAC/1.1a/ 2072014-2019)

remaining 60–70% is released in the form of faeces, urine and ammonia (FAO 2014). On average, 1 kg of feed (30% crude protein) globally releases about 27.6 g of N, and 1 kg of fish biomass releases about 577 g of BOD (biological oxygen demand), 90.4 g of N and 10.5 g of P (Tyson et al. 2011).

Aquaponics is currently a small but rapidly growing sector which is clearly suited to take advantage of the following political and socio-economic challenges, where 1) aquatic produce meets the need for food security and nutrition, 2) fish self-sufficient regions are established around the world, 3) aquaculture is a key sector but global ingredients and global feed production comes under focus, 4) innovation in agriculture promotes biodiversity in more sustainable ways and as part of the circular economy and 5) there is a greater take up of locally produced foods. These aspects tie in with the recommendations from the International Union for the Conservation of Nature (Le Gouvello et al. 2017), regarded the sustainability of the aquaculture and fish feed, which has recommended that efforts should be made to localize aquaculture production and the circular approach, and for putting in place a quality control programme for new products and by-products, as well as processing local fish feed within regions. So far, aquaponics as 'small-scale aquaculture farms' could provide examples for the implementation of the bioeconomy and local-scale production, thus promoting ways of using products and by-products from organic matter not suitable for use for other purposes, e.g. farmed insects and worms, macro- and microalgae, fish and by-product hydrolysates, new agro-ecology-produced plants and locally produced bioactives and micronutrients, whilst reducing the environmental footprint with quality food (fish and plants) production and moving towards zero waste generation. Moreover, aquaponics provides a good example for promoting a multidisciplinary way of learning about sustainable production and bioresource valorization, e.g. the 'Islandap Project' (INTERREG V-A MAC 2014-2020) (Fig.).

The following sections of this chapter review the state of the art of fish diets, ingredients and additives, as well as nutritional/sustainable challenges to consider when producing specific aquaponic feeds.

Sustainable Development of Fish Nutrition

The sustainable development of fish nutrition in aquaculture will need to correspond with the challenges that aquaponics delivers with respect to the growing need for producing high-quality food. Manipulating the nitrogen, phosphorus and the mineral content of fish diets used in aquaponics is one way of influencing the rates of the accumulation of nutrients, thereby reducing the need for the artificial and external supplementation of nutrients. According to Rakocy et al. (2004), fish and feed waste provide most of the nutrients required by plants if the optimum ratio between daily fish feed input and plant growing areas is sustained. Solid fish waste called 'sludge' in aquaponic systems results in losing approximately half of the available input nutrients, especially phosphorus, that theoretically could be used for plant biomass production but information is still limited (Delaide et al. 2017; Goddek et al. 2018). Whilst the goal of sustainability in fish nutrition in aquaculture will in the future be achieved by using tailor-made diets, fish feed in aquaponics needs to fulfil the nutritional requirements both for fish and for plants. Increases in sustainability will in part derive from less dependence on fishmeal (FM) and fish oil (FO) and novel, high-energy, low-carbon footprint raw natural ingredients. To safeguard biodiversity and the sustainable use of natural resources, the use of wild fisheries-based FM and FO needs to be limited in aquafeeds (Tacon and Metian 2015). However, fish performance, health and final product quality may be altered when substituting dietary FM with alternative ingredients. Thus, fish nutrition research is focused on the efficient use and transformation of the dietary components to provide the necessary essential nutrients that will maximize growth performance and achieve sustainable and resilient aquaculture. Replacing FM, which is an excellent but costly protein source in fish diets, is not straightforward due to its unique amino acid profile, high nutrient digestibility, high palatability, adequate amounts of micronutrients, as well as having a general lack of anti-nutritional factors (Gatlin et al. 2007).

Many studies have shown that FM can be successfully replaced by soybean meal in aquafeeds, but soybean meal has anti-nutritional factors such as trypsin inhibitors, soybean agglutinin and saponin, which limit its use and high replacement percentages in farming carnivorous fish. High FM replacement by plant meals in fish diets can also reduce nutrient bioavailability in fish, which results in nutrient alterations in the final quality of the product (Gatlin et al. 2007). It can also cause undesirable disturbances to the aquatic environment (Hardy 2010) and reduce fish growth due to the reduced levels of essential amino acids (especially methionine and lysine), and reduced palatability (Krogdahl et al. 2010). Gerile and Pirhonen (2017) noted that a 100% FM replacement with corn gluten meal significantly reduced growth rate of rainbow trout but FM replacement did not affect oxygen consumption or swimming capacity.

High levels of plant material can also affect the physical quality of the pellets, and may complicate the manufacturing process during extrusion. Most of the alternative plant-derived nutrient sources for fish feeds contain a wide variety of anti-nutritional factors that interfere with fish protein metabolism by impairing digestion and utilization, therefore leading to increased N release in the environment which can affect fish health and welfare. In addition, diets including high levels of phytic acid altered phosphorus and protein digestion that lead to high N and P release into the surrounding environment. Feed intake and palatability, nutrient digestibility and retention may vary according to fish species' tolerance and levels and can change the quantity and composition of the fish waste. Taking into consideration these results, fish diet formulations in aquaponics should investigate 'the tolerance' dietary levels of anti-nutritional factors (i.e. phytate) for different feed ingredients and for each fish species used in aquaponics and also the effects of the addition of minerals such as Zn and phosphate in the diets. It also should be noted that even if plant material is regarded as an ecologically sound option to replace FM in aquafeeds, plants need irrigation, and thus may induce ecological impacts in the form of their water and ecological footprints (Pahlow et al. 2015) from nutrient run-off from the fields.

Terrestrial animal by-products such as non-ruminant processed animal proteins (PAPs) derived from monogastric farmed animals (e.g. poultry, pork) that are fit for human consumption at the point of slaughter (Category 3 materials, EC regulation 142/2011; EC regulation 56/2013) could also replace FM and support the circular economy. They have higher protein content, more favourable amino acid profiles and fewer carbohydrates compared to plant feed ingredients whilst also lacking anti-nutritional factors (Hertrampf and Piedad-Pascual 2000). It has been shown that meat and bone meals may serve as a good phosphorus source when it is included in the diet of Nile tilapia (Ashraf et al. 2013), although it has been strictly banned in the feed of ruminant animals due to the danger of initiating bovine spongiform encephalopathy (mad cow disease). Certain insect species, such as black soldier fly (*Hermetia illucens*), could be used as an alternative protein source for sustainable fish feed diets. The major environmental advantages of insect farming are that (a) less land and water are required, (b) that greenhouse gas emissions are lower and that (c) insects have high feed conversion efficiencies (Henry et al. 2015). However, there is a continuing need for further research to provide evidence on quality and safety issues and screening for risks to fish, plants, people and the environment.

It is important to note that fish cannot synthesize several essential nutrients required for their metabolism and growth and depend on the feed for this supply. However, there are certain animal groups that can use nutrient-deficient diets, as they bear symbiotic microorganisms that can provide these compounds (Douglas 2010), and thus, fish can obtain maximal benefit when the microbial supply of their essential nutrients is scaled to demand. Undersupply limits fish growth, whilst oversupply can be harmful due to the need for the fish to neutralize toxicity caused by non-essential compounds. The extent to which the microbial function varies with the demands of different fish species and what are the underlying mechanisms are largely unknown. Importantly, an aquatic animal's gut microbiota can in theory play a critical role in providing the necessary nutrients and obtaining sustainability in fish farming (Kormas et al. 2014; Mente et al. 2016). Further research in this field will help facilitate the selection of ingredients to be used in fish feeds that promote gut microbiota diversity to improve fish growth and health.

Research into the utilization of alternative plant and animal protein sources and low trophic fish feed ingredients is ongoing. The substitution of marine sourced raw ingredients in fish feed, which could be used directly for human food purposes should decrease fishing pressure and contribute to preserving biodiversity. Low trophic-level organisms, such as black soldier fly, which may serve as aquafeed ingredients may be grown on by-products and waste of other agricultural industrial practices given different nutritional quality meals, thereby adding additional environmental benefits. However, efforts to succeed with the circular economy and the recycling of organic and inorganic nutrients should be handled with care since undesirable compounds in raw materials and seafood products could increase the risk to animal health, welfare, growth performance and safety of the final product for consumers. Research and continuous monitoring and reporting on contaminants of farmed aquatic animals in relation to the maximum limits in feed ingredients and diets are essential to inform revisions in and introductions of new regulations.

Feed Ingredients and Additives

Protein and Lipid Sources for Aquafeeds

Since the end of the twentieth century, there have been significant changes in the composition of aquafeeds but also advances in manufacturing. These transformations have originated from the need to improve the economic profitability of aquaculture as well as to mitigate its environmental impacts. However, the driving forces behind these changes is the need to decrease the amount of fishmeal (FM) and fish oil (FO) in the feeds, which have traditionally constituted the largest proportion of the feeds, especially for carnivorous fish and shrimp. Partly because of overfishing but especially due to the continuous increase in global aquaculture volume, there is

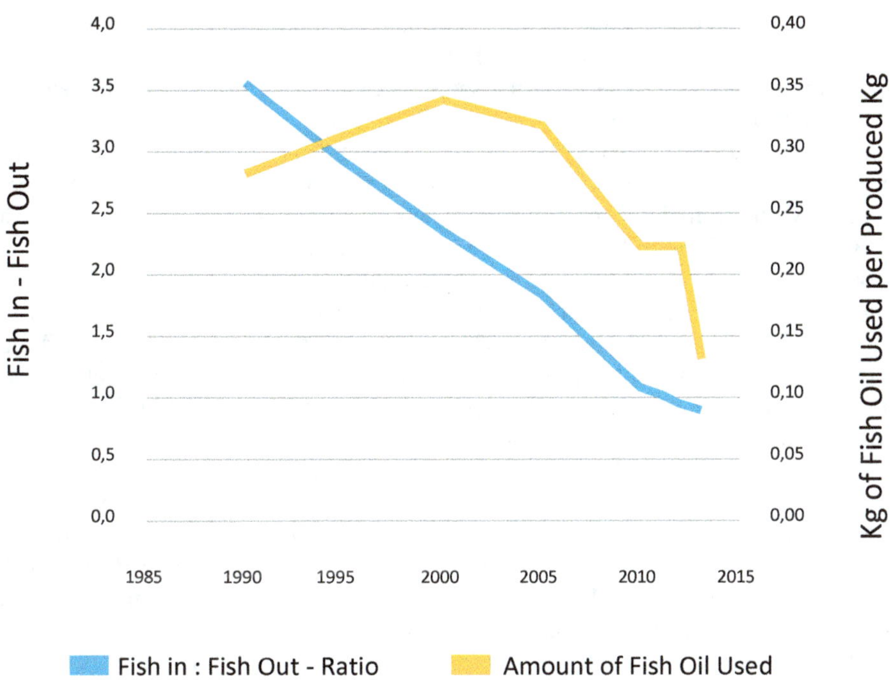

Fig. Fish-in-fish-out ratio (blue line, left y-axis) and the amount of fish oil used (yellow line, right y-axis) for rainbow trout feed in Finland between 1990 and 2013. (Data from www.raisioagro. com)

an increasing need for alternative proteins and oils to replace FM and FO in aquafeeds.

The composition of fish feeds has changed considerably as the proportion of FM in diets has decreased from >60% in the 1990s to <20% in modern diets for carnivorous fish such as Atlantic salmon (*Salmo salar*), and FO content has decreased from 24% to 10% (Ytrestøyl et al. 2015). As a consequence the so-called fish-in-fish-out (FIFO) ratio has decreased below 1 for salmon and rainbow trout meaning that the amount of fish needed in the feed to produce 1 kg of fish meat is less than 1 kg (Fig.). Thus, carnivorous fish culture in the twenty-first century is a net producer of fish. On the other hand, feeds for lower trophic omnivorous fish species (e.g. carp and tilapia) may contain less than 5% FM (Tacon et al. 2011). Farming such low trophic fish species is ecologically more sustainable than for higher trophic species, and FIFO for tilapia was 0.15 and for cyprinids (carp species) only 0.02 in 2015 (IFFO). It should be noted that total FM replacement in the diets of tilapia (Koch et al. 2016) and salmon (Davidson et al. 2018) is not possible without significantly affecting production parameters.

Today, the major supply of proteins and lipids in fish feed comes from plants, but also commonly from other sectors, including meals and fats from meat and poultry by-products and blood meal (Tacon and Metian 2008). Additionally, waste and by-products from fish processing (offal and waste trimmings) are commonly used to produce FM and FO. However, due to EU regulations (EC 2009), the use of FM of a species is not allowed as feed for the same species, e.g. salmon cannot be fed FM containing salmon trimmings.

FM and FO replacements with other ingredients can affect the quality of the product that is sold to customers. Fish has the reputation of being a healthy food, especially due to its high content of poly and highly unsaturated fatty acids. Most importantly, seafood is the only source of EPA (eicosapentaenoic acid) and DHA (docosahexaenoic acid), both of which are omega-3 fatty acids, and essential nutrients for many functions in the human body. If FM and FO are replaced with products from terrestrial origin, this will directly affect the quality of the fish flesh, most of all its fatty acid composition, as the proportion of omega-3 fatty acids (especially EPA and DHA) will decrease whilst the amount of omega-6 fatty acids will increase along with the increase of plant material that is replacing FM and FO (Lazzarotto et al. 2018). As such, the health benefits of fish consumption are partly lost, and the product that ends up on the plate is not necessarily what consumers expected to buy. However, in order to overcome the problem of decreased omega-3 fatty acids in the final product resulting from lower fish ingredients in aquafeeds, fish farmers could employ so-called finishing diets with high FO content during the final stages of cultivation (Suomela et al. 2017).

A new interesting option for replacing FO in fish feeds is the possibility of genetic engineering, i.e. genetically modified plants which can produce EPA and DHA, e.g. oil from genetically modified *Camelina sativa* (common name of camelina, gold-of-pleasure or false flax which is known to have high levels of omega-3 fatty acids) was successfully used to grow salmon, ending up with very high concentration on EPA and DHA in the fish (Betancor et al. 2017). The use of genetically modified organisms in human food production, however, is subject to regulatory approval and may not be an option in the short term.

Another new possibility to replace FM in aquafeeds is proteins made of insects (Makkar et al. 2014). This new option has become possible within the EU only recently when the EU changed legislation, allowing insect meals in aquafeeds (EU 2017). The species permitted to be used are black soldier fly (*Hermetia illucens*), common housefly (*Musca domestica*), yellow mealworm (*Tenebrio molitor*), lesser mealworm (*Alphitobius diaperinus*), house cricket (*Acheta domesticus*), banded cricket (*Gryllodes sigillatus*) and field cricket (*Gryllus assimilis*). Insects must be reared on certain permitted substrates. Growth experiments done with different fish species show that replacing FM with meal made of black soldier fly larvae does not necessarily compromise growth and other production parameters (Van Huis and Oonincx 2017). On the other hand, meals made of yellow mealworm could replace FM only partially to avoid decrease in growth (Van Huis and Oonincx 2017). However, FM replacement with insect meal can cause a drop in omega-3 fatty acids, as they are void of EPA and DHA (Makkar and Ankers 2014).

In contrast to insects, microalgae typically have nutritionally favourable amino acid and fatty acid (including EPA and DHA) profiles but there is also a wide variation between species in this respect. Partial replacement of FM and FO in aquafeeds with certain microalgae have given promising results (Camacho-Rodríguez et al. 2017; Shah et al. 2018) and in the future the use of microalgae in aquafeeds can be expected to increase (White 2017) even though their use may be limited by price.

This short summary of potential feed ingredients indicates that there is wide range of possibilities to at least partially replace FM and FO in fish feeds. In general, the amino acid profile of FM is optimal for most fish species and FO contains DHA and EPA which are practically impossible to provide from terrestrial oils, albeit genetic engineering may change the situation in the future. However, GMO products need first to be accepted in legislation and then by customers.

The Use of Specialist Feed Additives Tailored for Aquaponics

Tailoring aquafeeds which are specific to aquaponic systems is more challenging than conventional aquaculture feed development, as the nature of aquaponic systems requires that the aquafeeds not only supply nutrition to the cultured animals but also to the cultured plants and the microbial communities inhabiting the system. Current aquaponic practice utilizes aquafeeds formulated to provide optimal nutrition to the cultured aquatic animals; however, as the major nutrient input into aquaponic systems (Roosta and Hamidpour 2011; Tyson et al. 2011; Junge et al. 2017), the feeds also need to take into account the nutrient requirements of the plant production component. This is especially important for commercial-scale aquaponic systems, where the productivity of the plant production system has a major impact on overall system profitability (Adler et al. 2000; Palm et al. 2014; Love et al. 2015a) and where improved production performance of the plant component can significantly improve overall system profitability.

Thus, the overall aim of developing tailored aquaponic feeds would be to design a feed which strikes a balance between providing additional plant nutrients, whilst maintaining acceptable aquaponic system operation (i.e. sufficient water quality for animal production, biofilter and anaerobic digester performance, and nutrient absorption by plants). In order to achieve this, the final tailored aquaponic feed may not be optimal for either aquatic animal or plant production individually, but would rather be optimal for the aquaponic system as a whole. The optimal point would be determined based on overall system performance parameters, e.g. economic and/or environmental sustainability measures.

One of the major challenges in increasing production output from coupled aquaponic systems is the relatively low concentrations of both macro- and micro plant nutrients (mostly in the inorganic form) in the recirculating water, compared to conventional hydroponic systems. These low levels of nutrients can result in nutrient deficiencies in the plants and suboptimal plant production rates (Graber and Junge 2009; Kloas et al. 2015; Goddek et al. 2015; Bittsanszky et al.

2016; Delaide et al. 2017). A further challenge is the significant amounts of sodium chloride in conventional fishmeal-based aquafeeds and the potential accumulation of sodium in aquaponic systems (Treadwell et al. 2010). Different approaches can be developed to address these challenges such as technological solutions, e.g. decoupled aquaponic systems (Goddek et al. 2016) (also see Chapter), direct nutrient supplementation in the plant production system via foliar spray or addition to the recirculating water (Rakocy et al. 2006; Roosta and Hamidpour 2011) , or the culture of better salt-tolerant plant (see Chapter). A new approach is the development of tailored aquafeeds specifically for use in aquaponics.

In order to address plant nutrient shortages in aquaponics, tailored aquaponic feeds need to increase the amount of plant-available nutrients, either by increasing the concentrations of specific nutrients after excretion by the cultured animals, or by rendering the nutrients more bio-available after excretion and biotransformation, for rapid uptake by the plants. Achieving this increased nutrient excretion is, however, not as simple as supplementing increased amounts of the desired nutrients to the aquaculture diets, as there are many (often conflicting) factors that need to be considered in an integrated aquaponic system. For example, although optimal plant production will require increased concentrations of specific nutrients, certain minerals, e.g. certain forms of iron and selenium, can be toxic to fish even at low concentrations and would therefore have maximum allowable levels in the circulating water (Endut et al. 2011; Tacon 1987). Apart from total nutrient levels, the ratio between nutrients (e.g. the P:N ratio) is also important for plant production (Buzby and Lin 2014), and imbalances in the ratios between nutrients can lead to accumulation of certain nutrients in aquaponic systems (Kloas et al. 2015). Furthermore, even if an aquaponic feed increases plant nutrient levels, the overall system water quality and pH still needs to be maintained within acceptable limits to ensure acceptable animal production, efficient nutrient absorption by plant roots, optimal operation of biofilters and anaerobic digesters (Goddek et al. 2015b; Rakocy et al. 2006) and to avoid precipitation of certain important nutrients like phosphates, as this will render them unavailable to plants (Tyson et al. 2011). To achieve this overall balance is no mean feat, as there are complex interactions between the different forms of nitrogen in the system (NH_3, NH_4^+, NO_2^-, NO_3^-), the system pH and the assortment of metals and other ions present in the system (Tyson et al. 2011; Goddek et al. 2015; Bittsanszky et al. 2016).

Common Nutrient Shortages in Aquaponic Systems

Plants require a range of macro- and micronutrients for growth and development. Aquaponic systems are commonly deficient in the plant macronutrients potassium (K), phosphorus (P), iron (Fe), manganese (Mn) and sulphur (S) (Graber and Junge 2009; Roosta and Hamidpour 2011). Nitrogen (N) is present in different forms in aquaponic systems, and is excreted as part of the protein metabolism of the cultured aquatic animals (Rakocy et al. 2006; Roosta and Hamidpour 2011; Tyson et al. 2011) after which it enters the nitrogen cycle in the integrated aquaponic environment. (Nitrogen is discussed in detail in Chapter and is therefore excluded from the present discussion.)

The use of selected specialist aquaculture feed additives can contribute to the development of tailored aquafeeds specifically for aquaponics, by providing additional nutrients to the cultured aquatic animals and/or plants, or by adjusting the ratio of nutrients. Aquaculture feed additives are diverse, with a wide range of functions and mechanisms of working. Functions can be nutritive and non-nutritive, and the additives can be targeted towards action in the feeds or towards the physiological processes of the cultured aquatic animals (Encarnação 2016). For the purposes of this chapter, emphasis is on three specific types of additives which could assist the tailoring of aquaponic diets: (1) mineral supplements added directly to the feeds, (2) minerals that are added co-incidentally as part of additives that serve a non-mineral purpose and (3) additives which render minerals, which are already present in the feeds, more available to the cultured aquatic animals and/or plants in aquaponic systems.

1. *Direct mineral supplementation in aquaponic feeds*

Supplementing minerals directly in aquaculture diets used in aquaponic systems is one potential method to either increase the amount of minerals excreted by the cultured animals or to add specific minerals required by the plants in aquaponic systems. Minerals are routinely added in the form of mineral premixes to aquaculture diets, to supply the cultured aquatic animals with the essential elements required for growth and development (Ng et al. 2001; NRC 2011). Any minerals not absorbed by the fish during digestion are excreted, and if these are in the soluble (mostly ionic) form in the aquaponic system, these are available for plant uptake (Tyson et al. 2011; Goddek et al. 2015). It is unclear how feasible such an approach would be, as there is scant information about the efficacy of adding mineral supplements to aquafeeds for the purpose of enhancing aquaponic plant production. In general, mineral requirements and metabolism in aquaculture are poorly understood compared to terrestrial animal production, and the feasibility of this approach is therefore not well described. Potential advantages to this approach would be that it could prove to be a fairly simple intervention to improve overall system performance, it could allow supplementation of a wide range of nutrients, and it is likely to be relatively low cost. However, substantial research is still required to avoid any major potential pitfalls that may arise. One of these centres on the fact that the supplemented minerals destined for the plants first need to pass through the digestive tract of the cultured aquatic animals and these could either be absorbed fully or partially during this passage. This could lead to unwanted accumulation of minerals in the aquatic animals, or interference in normal intestinal nutrient and/or mineral absorption and physiological processes (Oliva-Teles 2012). Significant interactions can occur between dietary minerals in aquaculture diets (Davis and Gatlin 1996), and these need to be determined before direct mineral supplementation in aquaponic diets can be employed. Other potential effects may include altered physical structure and chemosensory characteristics of the feeds, which in turn could affect feed palatability. Clearly, there is still substantial research required before this method of tailoring aquaponic feeds can be adopted.

2. *Co-incidental addition of minerals by way of feed additives*

Certain classes of feed additives are added to aquafeeds in the form of ionic compounds and where only one of the ions contributes towards the intended activity. The other ion is viewed as a co-incidental and unavoidable addition to the aquafeed and is often not considered in any aquaculture research. One specific example of such a class of often-used feed additives are the organic acid salts, where the intended active ingredient in the aquafeed is the anion of an organic acid (e.g. formate, acetate, butyrate or lactate) and the accompanying cation is often ignored in the cultured animals' nutrition. Thus, if the accompanying cation is chosen purposefully to be an important macro- or micro plant nutrient, there is the potential that it could be excreted by the cultured animals into the system water and be available for uptake by the plants.

Short-chain organic acids and their salts have become well known and often used in feed additives in both terrestrial animal nutrition and in aquaculture, where the compounds are employed as performance enhancers and agents to improve disease resistance. These compounds can have different mechanisms of functioning, including acting as antimicrobials, antibiotics or growth promoters, enhancing nutrient digestibility and utilization and acting as directly metabolizable energy source (Partanen and Mroz 1999; Lückstädt 2008; Ng and Koh 2017). Either the native organic acids or their salts can be utilized in aquaculture diets, but the salt forms of the compounds are often preferred by manufacturers as they are less corrosive to feed manufacturing equipment, are less pungent and are available in solid (powder) form, which simplifies addition to formulated feed during manufacturing (Encarnação 2016; Ng and Koh 2017). For a comprehensive review on the use of organic acids and their salts in aquaculture, readers are referred to the work of Ng and Koh (2017).

Employing organic acid salts in aquaponics has the potential to have dual benefits in the system, where the anion could enhance the performance and disease resistance of the cultured aquatic animals, whilst the cation (e.g. potassium) could increase the amount of essential plant nutrients excreted. The potential advantage of this approach is that dietary inclusion levels of organic acid salts can be relatively high for a feed additive, and research regularly reports total organic acid salt inclusion of up to 2% by weight (Encarnação 2016), although commercial manufacturers tend to recommend lower levels of approximately 0.15–0.5% (Ng and Koh 2017). The cation of organic acid salts could constitute a significant proportion of the overall weight of the salt, and as these are fed daily to the cultured animals, they could contribute a significant amount of nutrients to the plants in an aquaponic system over the course of a growing season. No published research is currently available that reports findings for this line of enquiry and as with direct mineral supplementation to aquaponic feeds, this approach needs to be validated through future research to determine the fate of the cations added as part of the organic acid salts (whether they are excreted or absorbed by the aquatic animals), and whether there are any interactions with minerals or nutrients. It remains, however, an exciting future avenue of investigation.

3. *Feed additives that render nutrients more bio-available to plants*

Increasing amounts of plant ingredients are used in formulated aquafeeds, yet minerals from plant raw materials are less bio-available to cultured aquatic animals, mainly due to the presence of anti-nutritional factors in the plant-based dietary ingredients (Naylor et al. 2009; Kumar et al. 2012; Prabhu et al. 2016). This means that a higher proportion of minerals are excreted in the faeces in bound form, requiring 'liberation' before being available for plant uptake. One typical example is organic phosphorus occurring as phytate, which can bind to other minerals to form insoluble compounds, where microbial action in the environment is required before the phosphorus is released as plant-available, soluble phosphate (Kumar et al. 2012).

The use of exogenous enzymes in tailored aquaponic diets could potentially contribute towards releasing increased amounts of nutrients from high-plant content aquafeeds for both animal and plant nutrition in aquaponic systems. The most often employed enzymes in aquafeeds are proteases, carbohydrates and phytases, both to improve nutrient digestion and to degrade anti-nutritional compounds like phytate (Encarnação 2016), which can result in additional nutrients being released from aquafeeds. Although it is known that exogenous enzyme supplementation leads to improved nutrient utilization in the cultured animals, it is unclear whether additional nutrients would be excreted in plant-available form, therefore avoiding a separate remineralization step in aquaponic systems (see Chapter). Additionally, interactions between exogenous enzymes and nutrients in different parts of the digestive tract of fish are possible (Kumar et al. 2012), which will have further implications for the amounts of nutrients excreted for plant growth. Further research is therefore also required to determine the utility of exogenous enzymes specifically for use in aquaponic feeds.

Physiological Rhythms: Matching Fish and Plant Nutrition

The design of feeds for fish is crucial in aquaponics because fish feed is the single or at least the main input of nutrients for both animals (macronutrients) and plants (minerals) (Fig.).

Nitrogen is introduced to the aquaponic system through protein in fish feed which is metabolized by fish and excreted in the form of ammonia. The integration of recirculating aquaculture with hydroponics can reduce the discharge of unwanted nutrients to the environment as well as generate profits. In an early economic study, phosphorus removal in an integrated trout and lettuce/basil aquaponic system proved to be cost-saving (Adler et al. 2000). Integrating fish feeding rates is also paramount to fulfil the nutritional requirements of plants. Actually, farmers need to know the amount of feed used in the aquaculture unit to calculate how much nutrient needs to be supplemented to promote plant growth in the hydroponic unit. For instance, in a

Fig. Nutrient flow in an aquaponic system. Note that fish feed, through wastewater from the aquaculture system, provides the minerals required for plants to grow in the hydroponic system. Meal timing should be designed to match feeding/excretion rhythms in fish and nutrient uptake rhythms in plants

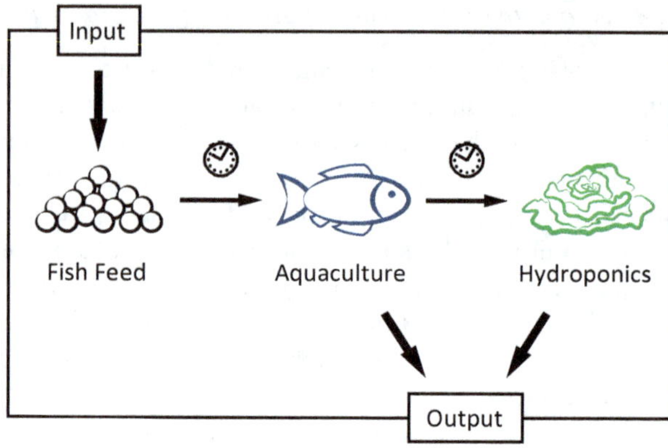

tilapia-strawberry aquaponic system, the total amount of feed required to produce ions (e.g. NO_3^{-}, Ca_2^{+}, $H_2PO_4^{-}$ and K^{+}) for plants was calculated at different fish densities, with better result for small fish density 2 kg fish/m^3 to reduce the cost of hydroponic solution supplementation (Villarroel et al. 2011).

It is well known that plants have daily rhythms and circadian rhythmicity in leaf movements was first described in plants by de Mairan in the early eighteenth century (McClung 2006). Circadian rhythms in plants control everything from the time of flowering to plant nutrition and thus these rhythms need to be taken into account especially when using artificial horticultural lighting. Fish are also tied into daily rhythms in most physiological functions, including feeding and nutrient uptake. It should not be surprising that fish exhibit feeding rhythms because food availability and the occurrence of predators are hardly constant but restricted to a particular time of day/night (López-Olmeda and Sánchez-Vázquez 2010). Thus, fish should be fed at the right time according to their appetite rhythms: meals scheduled during daytime for diurnal fish species, and at night for nocturnal fish. It is well known that fish show daily patterns of deamination of proteins and nitrogenous wastes related to their nutritional status and feeding rhythms (Kaushik 1980). Feeding time affects nitrogen excretion, as Gelineau et al. (1998) reported ammonia production and protein catabolism were lower in fish fed at dawn (in phase with their feeding rhythm) than in those fed at midnight (out of phase). Most interestingly, urea excretion shows circadian rhythmicity that persist in starved fish under constant conditions (Kajimura et al. 2002), revealing its endogenous origin. Furthermore, urea permeability (determined as body urea contents after immersion in a urea solution) coincided with the acrophase, i.e. the peak of the daily excretion rhythm, indicating that urea does not permeate cells by simple diffusion but there is a circadian control. Plants also show daily rhythms in nitrogen uptake, as early described by Pearson and Steer (1977), who found a daily pattern of nitrate uptake and nitrate reductase in peppers kept in a constant environment. The nitrate concentration in the leaves of spinach also increased during the night as the uptake rate of nitrate by the roots increase at that time (Steingrover et al. 1986). In aquaponics, the evidence thus points to the need for matching excretion rhythms in fish and nutrient uptake rhythms in plants. To optimize the performance and cost-effectiveness of aquaponic systems, fish diets and feeding schedules should be designed carefully to provide nutrients at the right level and the right time to complement both fish and plants.

References

Adler PR, Harper JK, Wade EM, Takeda F, Summerfelt ST (2000) Economic analysis of an aquaponic system for the integrated production of rainbow trout and plants. Int J Recirculat Aquacult 1:15–34

Alltech (2017) Alltech yearly survey 2017. https://go.alltech.com/alltech-feed-survey-2017

Ashraf S, Rania SM, Ehab REH (2013) Meat and bone meal as a potential source of phosphorus in plant-protein-based diets for Nile tilapia (*Oreochromis niloticus*). Aquacult Intl 21:375–385

Béné C, Barange M, Subasinghe R, Pinstrup-Andersen P, Merino G, Hemre G-I, Williams M (2015) Feeding 9 billion by 2050 – putting fish back on the menú. Food Sec 7:261–274. https://doi.org/10.1007/s12571-015-0427-z

Betancor MB, Li K, Sprague M, Bardal T, Sayanova O, Usher S, Han L, Måsøval K, Torrissen O, Napier JA, Tocher DR, Olsen RO (2017) An oil containing EPA and DHA from transgenic *Camelina sativa* to replace marine fish oil in feeds for Atlantic salmon (*Salmo salar* L.): Effects on intestinal transcriptome, histology, tissue fatty acid profiles and plasma biochemistry. PLoS ONE 12(4):e0175415. https://doi.org/10.1371/journal.pone.0175415

Beveridge MCM, Thilsted SH, Phillips MJ, Metian M, Troell M, Hall SJ (2013) Meeting the food and nutrition needs of the poor. J Fish Biol:1067. https://doi.org/10.1111/jfb.12187

Bittsanszky A, Uzinger N, Gyulai G, Mathis A, Junge R, Villarroel M, Kotzen B, Komives T (2016) Nutrient supply of plants in aquaponic systems. Ecocycles 2:17–20

Buzby KM, Lin L-S (2014) Scaling aquaponic systems: Balancing plant uptake with fish output. Aquacult Eng 63:39–44. https://doi.org/10.1016/j.aquaeng.2014.09.002.

Camacho-Rodríguez J, Macías-Sánchez MD, Cerón-García MC, Alarcón FJ, Molina-Grima E (2017) Microalgae as a potential ingredient for partial fish meal replacement in aquafeeds: nutrient stability under different storage conditions. J Appl Phycol. https://doi.org/10.1007/s10811-017-1281-5

Davidson J, Kenney PB, Barrows FT, Good C, Summerfelt ST (2018) Fillet quality and processing attributes of post-smolt Atlantic salmon, *Salmo salar*, fed a fishmeal-free diet and a fishmeal-based diet in recirculation aquaculture systems. J World Aquacult Soc 49:183–196. https://doi.org/10.1111/jwas.12452

Davis DA, Gatlin DM (1996) Dietary mineral requirement of fish and marine crustaceans. Rev Fish Sci 4:75–99. https://doi.org/10.1080/10641269609388579

Delaide B, Delhaye G, Dermience M, Gott J, Soyeurt H, Jijakli MH (2017) Plant and fish production performance, nutrient mass balances, energy and water use of the PAFF Box, a small-scale aquaponic system. Aquac Eng 78:130–139

Douglas AE (2010) The symbiotic habit. Princeton University Press, Princeton

Ebeling JM, Timmons MB (2012) Recirculating aquaculture systems. In: Tidwell JH (ed) Aquaculture production systems. Wiley, Hoboken

EC (2009) Regulation (EC) No 1069/2009 of the European parliament and of the council of 21 October 2009 laying down health rules as regards animal by-products and derived products not intended for human consumption and repealing regulation (EC) No 1774/2002 (animal by-products regulation). Off J Eur Union L 300/1. https://doi.org/10.3000/17252555.L_2009.300.eng

Encarnação P (2016) Chapter: Functional feed additives in aquaculture feeds. In: Nates SF (ed) Aquafeed formulation. Academic, San Diego, pp 217–237

Endut A, Jusoh A, Ali N, Wan Nik WB (2011) Nutrient removal from aquaculture wastewater by vegetable production in aquaponics recirculation system. Desalinat Water Treat 32:422–430. https://doi.org/10.5004/dwt.2011.2761

EU (2017) Commission Regulation (EU) 2017/893 of 24 May 2017 amending Annexes I and IV to Regulation (EC) No 999/2001 of the European Parliament and of the Council and Annexes X, XIV and XV to Commission Regulation (EU) No 142/2011 as regards the provisions on processed animal protein. Off J Eur Union L138/92. http://data.europa.eu/eli/reg/2017/893/oj

FAO (2014) The State of World Fisheries and Aquaculture. Food and Agricultural Organization, Rome, Italy

FAOSTAT (2015) Fish and fishery products – world apparent consumption statistics based on food balance sheets (1961–)

García-Romero J, Gines R, Izquierdo M, Robaina L (2014a) Marine and freshwater crab meals in diet for red porgy (*Pagrus pagrus*): effect on fillet fatty acid profile and flesh quality. Aquaculture 420–421:231–239. https://doi.org/10.1016/j.aquaculture.2013.10.035

García-Romero J, Gines R, Vargas R, Izquierdo M, Robaina L (2014b) Marine and freshwater crab meals in diet for red porgy (*Pagrus pagrus*): Digestibility, ammonia-N excretion, phosphorus and calcium retention. *Aquaculture* 428–429:158–165. https://doi.org/10.1016/j.aquaculture.2014.02.035

Gatlin DM, Barrows FT, Brown P, Dabrowski K, Gaylord TG, Hardy RW, Herman E, Hu GS, Krogdahl A, Nelson R, Rust M, Sealey W, Skonberg D, Souza EJ, Stone D, Wilson R, Wurtele E (2007) Expanding the utilization of sustainable plant products in aquafeeds; A review. Aquacult Res 38:551–579

Gelineau A, Medale F, Boujard T (1998) Effect of feeding time on postprandial nitrogen excretion and energy expenditure in rainbow trout. J Fish Biol 52:655–664

Gerile S, Pirhonen J (2017) Replacement of fishmeal with corn gluten meal in feeds for juvenile rainbow trout (*Oncorhynchus mykiss*) does not affect oxygen consumption during forced swimming. Aquaculture 479:616–618. https://doi.org/10.1016/j.aquaculture.2017.07.002

Goddek S, Delaide B, Mankasingh U, Ragnarsdottir KV, Jijakli H, Thorarinsdottir R (2015) Challenges of sustainable and commercial aquaponics. Sustainability 7:4199–4224. https://doi.org/10.3390/su7044199.

Goddek S, Espinal CA, Delaide B, Jijakli MH, Schmautz Z, Wuertz S, Keesman KJ (2016) Navigating towards decoupled Aquaponic systems: a system dynamics design approach. Water 8:303. https://doi.org/10.13140/RG.2.1.3930.0246

Goddek S, Delaide BPL, Joyce A, Wuertz S, Jijakli HM, Grosse A, Eding EH, Bläser I, Reuterg M, Keizer LCP, Morgenstern R, Körner O, Verreth J, Keesman KJ (2018) Nutrient mineralization and organic matter reduction performance of RAS-based sludge in sequential UASB-EGSB reactors. Aquacult Eng 83:10–19. https://doi.org/10.1016/j.aquaeng.2018.07.003

Graber A, Junge R (2009) Aquaponic systems: nutrient recycling from fish wastewater by vegetable production. Desalination 246:147–156

Hardy RW (2010) Utilization of plant proteins in fish diets: effects of global demand and supplies of fishmeal. Review article. Aquacult Res 41:770–776

Henry M, Gasco L, Piccolo G, Fountoulaki E (2015) Review on the use of insects in the diet of farmed fish: past and future. Anim Feed Sci Technol 203:1–22

Hertrampf JW, Piedad-Pascual F (2000) Handbook on ingredients for aquaculture feeds. Kluwer Academic Publishers, Dordrecht. 624 pp

IFFO, The Marine Ingredients Association. http://www.iffo.net

Junge R, König B, Villarroel M, Komives T, Jijakli H (2017) Strategic points in aquaponics. Water 9(3):182. https://doi.org/10.3390/w9030182

Kajimura M, Iwata K, Numata H (2002) Diurnal nitrogen excretion rhythm of the functionally ureogenic gobiid fish *Mugilogobius abei*. Comp Biochem Physiol B 131:227–239

Kaushik SJ (1980) Influence of the nutritional status on the daily pattern of nitrogen excretion in the carp (*Cyprinus carpio* L.) and the rainbow trout (*Salmo gairdneri* R.). Reprod Nut Develop 20:1751–1765

Kaushik S (2017) Aquaculture deals with the production of all kinds of aquatic organisms through human intervention, 2017 meeting. International Council of Academies of Engineering and Technological Sciences (CAETS – http://www.caets.org). Madrid, 14–15 November 2017

Khakyzadeh V, Luque R, Zolfigol MA, Vahidian HR, Salehzadeh H, Moradi V, Soleymani AR, Moosavi-Zare AR, Xu K (2015) Waste to wealth: a sustainable aquaponic system based on residual nitrogen photoconversion. Royal Society of Chemistry 5:3917–3921. https://doi.org/10.1039/C4RA15242E

Kingler D, Naylor R (2012) Searching for Solutions in Aquaculture: Charting a Sustainable Course. Anual Rev Environ Resour 37:247–276

Kloas W, Groß R, Baganz D, Graupner J, Monsees H, Schmidt U, Staaks G, Suhl J, Tschirner M, Wittstock B, Wuertz S, Zikova A, Rennert B (2015) A new concept for aquaponic systems to improve sustainability, increase productivity, and reduce environmental impacts. Aquacult Environ Interact 7:179–192. https://doi.org/10.3354/aei00146

Koch JF, Rawlesb SD, Webster CD, Cummins V, Kobayashic Y, Thompson KR, Gannam AL, Twibell RG, Hyded NM (2016) Optimizing fish meal-free commercial diets for Nile tilapia, *Oreochromis niloticus*. Aquaculture 452:357–366. https://doi.org/10.1016/j.aquaculture.2015.11.017

Kormas KA, Meziti A, Mente E, Fretzos A (2014) Dietary differences are reflected on the gut prokaryotic community structure of wild and commercially reared sea bream (*Sparus aurata*). Microbiology open. https://doi.org/10.1002/mbo3.202

Kobayashi M, Msangi S, Batka M, Vannuccini S, Dey MM, Anderson JL (2015) Fish to 2030: the role and opportunity for aquaculture. Aquacult Econ Manage 19:282–300. https://doi.org/10.1080/13657305.2015.994240

Krogdahl A, Penn M, Thorsen J, Refstie S, Bakke AM (2010) Important anti-nutrients in plant feedstuffs for aquaculture: An update on recent findings regarding responses in salmonids. Aquacult Res 41:333–344

Kumar V, Sinha AK, Makkar HPS, De Boeck G, Becker K (2012) Phytate and phytase in fish nutrition. J Anim Physiol Anim Nutr 96:335–364. https://doi.org/10.1111/j.1439-0396.2011.01169.x

Lazzarotto, V., Médale, F., Larroquet, L. & Corraze, G. (2018). Long-term dietary replacement of fishmeal and fish oil in diets for rainbow trout (*Oncorhynchus mykiss*): Effects on growth, whole body fatty acids and intestinal and hepatic gene expression. PLoS One 13(1) https://doi.org/10.1371/journal.pone.0190730

Le Gouvello, Raphaëla et François Simard (eds) (2017). Durabilité des alimments pour le poisson en aquaculture: Réflexions et recommandations sur les aspects technologiques, économique sociaux et environnementaux. Gland, Suisse: UICN, et Paris, France : Comité français de l'UICN. 296 pp

López-Olmeda JF, Sánchez-Vázquez FJ (2010) Feeding rhythms in fish: from behavioural to molecular approach. In: Kulczykowska E, Popek W. Kapoor BG (eds) Biological clock in fish. CRC Press, Enfield, pp 155–184

Love DC, Fry JP, Genello L, Hill ES, Frederick A, Li X, Semmens K (2014) An International survey of aquaponics practitioners. PLoS One 9(7):e102662. https://doi.org/10.1371/journal.pone.0102662

Love DC, Fry JP, Li X, Hill ES, Genello L, Semmens K, Thompson RE (2015a) Commercial aquaponics production and profitability: findings from an international survey. Aquaculture 435:67–74. https://doi.org/10.1016/j.aquaculture.2014.09.023

Love DC, Uhl MS, Genello L (2015b) Energy and water use of a small-scale raft aquaponics system in Baltimore, Maryland, United States. Aquacult Eng 68:19–27

Lückstädt C (2008) The use of acidifiers in fish nutrition. CAB Rev Perspect Agricult Veterinary Sci Nutr Natur Resour 3:1–8

Makkar HPS, Ankers P (2014) Towards sustainable animal diets: a survey-based study. Anim Feed Sci Technol 198:309–322. https://doi.org/10.1016/j.anifeedsci.2014.09.018

Makkar HPS, Tran G, Heuzé V, Ankers P (2014) State-of-the-art on use of insects as animal feed. Anim Feed Sci Techn 197:1–33. https://doi.org/10.1016/j.anifeedsci.2014.07.008

McClung CR (2006) Plant circadian rhythms. Plant Cell 18:792–803

Mente E, Gannon AT, Nikouli E, Hammer H, Kormas KA (2016) Gut microbial communities associated with the molting stages of the giant freshwater prawn *Macrobrachium rosenbergii*. Aquaculture 463:181–188

Msangi S, Kobayashi M, Batka M, Vannuccini S, Dey MM, Anderson JL (2013) Fish to 2030: prospects for fisheries and aquaculture. World Bank Report Number 83177-GLB. http://documents.worldbank.org/curated/en/458631468152376668/

Naylor RL, Hardy RW, Bureau DP, Chiu A, Elliott M, Farrell AP, Forster I, Gatlin DM, Goldburg RJ, Hua K, Nichols PD (2009) Feeding aquaculture in an era of finite resources. Proc Natl Acad Sci USA 106:15103–15110

Ng W-K, Koh C-B (2017) The utilization and mode of action of organic acids in the feeds of cultured aquatic animals. Rev Aquacult 9:342–368. https://doi.org/10.1111/raq.12141

Ng W-K, Ang L-P, Liew F-L (2001) An evaluation of mineral supplementation of fish meal-based diets for African catfish. Aquacult Int 9:277–282

NRC. N.R.C (2011) Nutrient requirements of fish and shrimp. The National Academies Press, Washington, District Columbia

Oliva-Teles A (2012) Nutrition and health of aquaculture fish. J Fish Dis 35:83–108. https://doi.org/10.1111/j.1365-2761.2011.01333.x

Pahlow M, Oel PR, Mekonnen MM, Hoekstra AY (2015) Increasing pressure on freshwater resources due to terrestrial feed ingredients for aquaculture production. Sci Total Environ 536:847–857. https://doi.org/10.1016/j.scitotenv.2015.07.124

Palm HW, Seidemann R, Wehofsky S, Knaus U (2014) Significant factors affecting the economic sustainability of closed aquaponic system. Part I: system design, chemo-physical parameters and general aspects. AACL Bioflux 7:20–32

Partanen KH, Mroz Z (1999) Organic acids for performance enhancement in pig diets. Nutr Res Rev 12:117–145

Pearson CJ, Steer BT (1977) Daily changes in nitrate uptake and metabolism in *Capsicum annuum*. Planta 137(2):107–112. https://doi.org/10.1007/BF00387546

Prabhu PAJ, Schrama JW, Kaushik SJ (2016) Mineral requirements of fish: a systematic review. Rev Aquacult 8:172–219

Rakocy JE, Shultz RC, Bailey DS, Thoman ES (2004) Aquaponic production of tilapia and basil: comparing a batch and staggered cropping system. Acta Hortic 648:63–69. https://doi.org/10.17660/ActaHortic.2004.648.8

Rakocy JE, Masser MP, Losordo TM (2006) Recirculating aquaculture tank production systems: aquaponics- integrating fish and plant culture. Southern Regional Aquaculture Center, pp 1–16

Robaina L, Izquierdo MS, Moyano FJ, Socorro J, Vergara JM, Montero D (1998) Increase of the dietary n-3/n-6 fatty acid ratio and addition of phosphorus improves liver histological alterations induced by feeding diets containing soybean meal to gilthead seabream, *Sparus aurata*. Aquaculture 161:281–293

Robaina L, Corraze G, Aguirre P, Blanc D, Melcion JP, Kaushik S (1999) Digestibility, postprandial ammonia excretion and selected plasma metabolites in European sea bass (*Dicentrarchus labrax*) fed pelleted or extruded diets with or without wheat gluten. Aquaculture 179:45–56

Roosta HR, Hamidpour M (2011) Effects of foliar application of some macro- and micro-nutrients on tomato plants in aquaponic and hydroponic systems. Scientia Horticulturae 129:396–402. https://doi.org/10.1016/j.scienta.2011.04.006

Shah MR, Giovanni Antonio Lutzu GA, Alam A, Sarker P, Chowdhury MAK, Parsaeimehr A, Liang Y, Daroch M (2018) Microalgae in aquafeeds for a sustainable aquaculture industry. J Appl Phycol 30:197–213. https://doi.org/10.1007/s10811-017-1234-z

Staples D, Funge-Smith S (2009) Ecosystem approach to fisheries and aquaculture: Implementing the FAO Code of Conduct for Responsible Fisheries. FAO Regional Office for Asia and the Pacific, Bangkok, Thailand. RAP Publication 2009/11, 48 pp

Steingrover E, Ratering P, Siesling J (1986) Daily changes in uptake, reduction and storage of nitrate in spinach grown at low light intensity. Physiol Plantarum 66:555–556

Suomela JP, Tarvainen M, Kallio H, Airaksinen S (2017) Fish oil finishing diet maintains optimal n-3 long-chain fatty acid content in European whitefish (*Coregonus lavaretus*). Lipids 52:849–855. https://doi.org/10.1007/s11745-017-4290-x

Tacon AGJ (1987) The nutrition and feeding of farmed fish and shrimp - a training manual. Food and Agricultural Organisation of the United Nations, Rome, Italy

Tacon AGJ, Metian M (2008) Global overview on the use of fish meal and fish oil in industrially compounded aquafeeds: trends and future prospects. Aquaculture 285:146–158. https://doi.org/10.1016/j.aquaculture.2008.08.015

Tacon AGJ, Metian M (2015) Feed matters: Satisfying the feed demand of aquaculture. Rev Fish Sci Aquacult 23(1):1–10. https://doi.org/10.1080/23308249.2014.987209

Tacon AGJ, Hasan MR, Metian M (2011) Demand and supply of feed ingredients for farmed fish and crustaceans: trends and prospects. FAO Fisheries and Aquaculture Technical Paper no. 564. FAO, Rome, 87 pp

Terova G, Robaina LE, Izquierdo MS, Cattaneo AG, Molinari S, Bernardini G, Saroglia M (2013) PepT1 mRNA expression levels in sea bream (*Sparus aurata*) fed different plant protein sources. Springer Plus 2:17. https://doi.org/10.1186/2193-1801-2-17

Thilsted SH, Thorne-Lyman A, Webb P, Bogard JR, Subasinghe R, Phillips MJ, Allison EH (2016) Sustaining healthy diets: The role of capture fisheries and aquaculture for improving nutrition in the post-2015 era. Food Policy 61:126–131. https://doi.org/10.1016/j.foodpol.2016.02.005

Torrecillas S, Robaina L, Caballero MJ, Montero D, Calandra G, Mompel D, Karalazos V, Sadasivam K, Izquierdo M (2017) Combined replacement of fishmeal and fish oil in European sea bass (*Dicentrarchus labrax*): production performance, tissue composition and liver morphology. Aquaculture 474. https://doi.org/10.1016/j.aquaculture.2017.03.031

Treadwell D, Taber S, Tyson R, Simonne E (2010) HS1163: Foliar-applied micronutrients in aquaponics: s guide to use and sourcing, IFAS Extension. University of Florida

Tveterås S, Asche F, Bellemare MF, Smith MD, Guttormsen AG, Lem A, Lien K, Vannuccini S (2012) Fish is food – the FAO's fish price index. PLoS ONE 7:e36731

Tyson RV, Treadwell DD, Simonne EH (2011) Opportunities and challenges to sustainability in aquaponic systems. HortTechnology 21:6–13

Van Huis A, Oonincx DGAB (2017) The environmental sustainability of insects as food and feed. A review. Agron Sustain Dev 37:43. https://doi.org/10.1007/s13593-017-0452-8

Villarroel M, Alvariño JMR, Duran JM (2011) Aquaponics: integrating fish feeding rates and ion waste production for strawberry hydroponics. Spanish J Agricult Res 9:537–545

White C (2017) Algae-based aquafeed firms breaking down barriers for fish-free feeds. https://www.seafoodsource.com/news/aquaculture/algae-based-aquafeed-firms-breaking-down-barriers-for-fish-free-feeds

Ytrestøy T, Aas TS, Åsgård T (2015) Utilisation of feed resources in production of Atlantic salmon (*Salmo salar*) in Norway. Aquaculture 448:365–374. https://doi.org/10.1016/j.aquaculture.2015.06.023

5

Commercial Aquaponics: A Long Road Ahead

Maja Turnšek, Rolf Morgenstern, Iris Schröter, Marcus Mergenthaler, Silke Hüttel, and Michael Leyer

Abstract Aquaponic systems are often designated as sustainable food production systems that are still facing various challenges, especially when they are considered as a commercial endeavour that needs to compete on the market. The early stages of the aquaponics industry have witnessed a number of unrealistic statements about the economic advantageousness of aquaponics. This chapter deals with these topics and discusses them critically. The latest scientific literature and current personal experiences of European commercial aquaponics farmers are taken into account on three levels: The horticulture side of production, the aquaculture side of production and the early data on the market response to aquaponics, emphasising the marketing issues and public acceptance of aquaponics. In summary, the chapter does not provide an "off-the-peg" solution to evaluate the economic performance of a particular aquaponics system. Instead it provides a broad database that enables an estimation of the efficiency of a planned system more realistically, pointing to challenges that the commercial aquaponics early adopters faced that are important lessons for future aquaponic endeavours, particularly in Europe.

Authors Maja Turnšek and Rolf Morgenstern have equally contributed to this chapter.

M. Turnšek (✉)
Faculty of Tourism, University of Maribor, Brežice, Slovenia
e-mail: maja.turnsek@um.si

R. Morgenstern · I. Schröter · M. Mergenthaler
Department of Agriculture, University of Applied Sciences of South Westphalia, Soest, Germany
e-mail: morgenstern.rolf@fh-swf.de; schroeter.iris@fh-swf.de; mergenthaler.marcus@fh-swf.de

S. Hüttel
Institute of Food and Resource Economics, Chair of Production Economics, Faculty of Agriculture, Rheinische Friedrichs-Wilhelms Universität Bonn, Bonn, Germany
e-mail: S.Huettel@ilr.uni-bonn.de

M. Leyer
Institute of Business Administration, Rostock University, Rostock, Germany
e-mail: michael.leyer@uni-rostock.de

Keywords Commercial aquaponics · Aquaponics economic myths · Marketing of aquaponics

Introduction: Beyond Myths

Although we have witnessed the first research developments in aquaponics as far back as the late 1970s (Naegel 1977; Lewis et al. 1978), there is still a long road ahead for the sound economical assessment of aquaponics. The industry is developing slowly, and thus available data is often based on model cases from research and not on commercial-based systems. After initial positive conclusions about the economic potentials of aquaponics in research-based settings of the low-investment systems in USA, primarily the system in Virgin Islands (Bailey et al. 1997) and Alberta (Savidov and Brooks 2004), commercial aquaponics encountered a high level of early enthusiasm in business contexts, often based on unrealistic expectations.

To provide a specific example, in its early market forecast, IndustryARC (2012) anticipated that aquaponics as an industry has a potential market size of around $180 million in 2013 and is expected to reach $1 billion in sales in 2020. Later they projected aquaponics to increase from $409 million in 2015 to $906.9 million by 2021 (IndustryARC 2017). The same report (IndustryARC 2012) provided a number of yet untested claims about aquaponics, for example, about the economic superiority of aquaponics in terms of output, growth time and diversification possibilities in a commercial setting. We name such claims here as "aquaponics economic myths" that have been a typical part of the early internet-fuelled hype on commercial aquaponics.

Take a look at their statement: "Aquaponics uses 90% less land and water than agriculture but has the potential to generate 3 to 4 times more food than the latter also" (IndustryARC 2012). Comments such as these are extremely vague, since it is not clear what exactly aquaponics is being compared to when the authors are referring to "agriculture". Although aquaponics does use less water than soil-based food production, since the water used in soil-based production can be lost in the soil, not being absorbed by plants compared to staying in a recirculation loop with aquaponics. The exact amount of water savings depends on the type of the system. Additionally, "3 to 4 times more food" seems highly exaggerated. Aquaponics can have yields comparable to hydroponics (e.g. Savidov and Brooks 2004; Graber and Junge, 2009). Yet the statement glosses over the fact that at least in coupled aquaponics so-called operational compromises need to be made in order to find a balance between optimum parameters for healthy plants and fish, which can lead to aquaponics having lower outputs compared to hydroponics.

Therefore, statements like the above lack a clear definition of the reference scenario and the reference unit of comparison. In an economic assessment, higher output levels can only be compared meaningfully if there is a clear reference to the input levels required to achieve this output. In the assessment of aquaponic systems, higher outputs per area might be achieved compared to conventional agriculture, yet aquaponic systems might require more energy, capital and work input. Only referring to land as an input factor assumes that other production factors are not scarce, which is hardly the case. Therefore, statements like the above neglect the "all other things being equal" principle in economic assessments. Vaclav Smil (2008) calculates and summarises energy expenditure of different agricultural activities, utilising energy as the common denominator, and this allows us to compare different agricultural methods with the aquaponic approach.

A similar myth is contained in the statement: "A major advantage pertaining to the aquaponics industry is that crop production time can be accelerated" (IndustryARC 2012). An acceleration of crop production necessarily depends on the amount of nutrients and water, oxygen and carbon dioxide in the surrounding atmosphere and light and temperature available to crops – factors that are not elements of aquaponics per se but can be added via greenhouse management practices, such as fertilisation and irrigation heating and artificial lights. These additional elements, however, increase both the costs of investment and the operational costs, often being too expensive to be economically viable (depending of course on the location, type of crops and especially the price of crops).

Another economically important advantage of aquaponics provided in the report was that "aquaponics is an adaptable process that allows for a diversification of income streams. Crops may be produced depending upon local market interest and the interest of the grower" (IndustryARC 2012). What statements like this gloss over is the fact that diversification of production always comes at a price. Crop diversification necessarily includes higher levels of knowledge and higher labour demands. The larger the variety of crops, the more difficult it is to meet optimum conditions for all the selected crops. Large-scale commercial production thus looks for constant parameters for a limited number of crops that need similar growth conditions, allowing for large outputs in order to penetrate distribution via large distribution partners such as supermarket chains, and allow for the same storage and potential processing equipment and processes. Such large-scale production is able to use economies of scale to reduce unit costs, a basic principle in economic assessment, which is not usually the case for aquaponics at smaller scales of production.

Finally, the most important statement provided in the report was that "the return of investment (ROI) for aquaponic systems ranges from 1 to 2 years depending on the farmer experience as well as the scale of farming" (IndustryARC 2012). Such statements need to be taken with extreme caution. The scarce data that is available on return on investments reports on a much longer time: According to Adler et al. (2000), it takes 7.5 years of return for an approximately $ 300.000 investment in the hypothetical scenario of a rainbow trout and lettuce system. Recently, Quagrainie et al. (2018) reported a similar period of 6.8 years for the payback of an investment in aquaponics if the products can only be sold at non-organic prices. Real data on the economics of aquaponics is extremely hard to come by, since the enterprises that have ventured into commercial aquaponics are reluctant to share their data. In cases where enterprises are performing well, they either do not share their data, since it is considered a business secret, or if they do share data, such data needs to be taken with caution since typically these companies have an interest in selling aquaponic equipment, engineering and consultancy. In addition, enterprises that have failed in achieving profitability prefer not to publicly share their failures.

These "myths" are continually circulating online amongst non-experienced aquaponic enthusiasts, fuelled by hope for both high returns and a path towards more sustainable future food production. So there is a need to go beyond the myths and look at individual enterprises and provide an in-depth analysis on the basic and the general economics of aquaponics.

Even if realistic data on aquaponics were available, it has to be considered that such analyses are based on single cases. As aquaponic systems are far from technically standardised production systems, the diversity with respect to marketing concepts is even higher. So, data on every single aquaponics system lacks generalisability and can be regarded only as a single case study. General statements are therefore not valid if the framework conditions and technical and marketing specificities are not made transparent.

Journalistic publications about aquaponics often follow a narrative that elaborates on the general challenges of global agriculture, such as shrinking agricultural areas, humus loss and desertification, and then elaborate on the advantages of aquaponic food production methods. Apart from the above-mentioned mistake that in fact controlled environment system (CES) production is compared with field production, no distinction between agriculture and horticulture is made. Whilst the term "agriculture" technically includes horticulture, agriculture in its more specific sense is the large-scale crop production on farmland. Horticulture is the cultivation of plants, usually excluding large-scale crop production on farmland, and is typically carried out in greenhouses. Following these definitions, the plant side of aquaponics is horticulture and not agriculture. Thus comparing yield and other productivity properties of aquaponics with agriculture is simply comparing apples to oranges.

To state this differently, the horticulture side of agriculture is only a very small part of it. Large-scale crop production in agriculture mostly encompasses so-called staple food production: Cereals like corn, barley and wheat, oilseed like rape and sunflower and starchy root vegetables like potatoes. The agricultural area of Germany covers 184.332 km^2 (Destatis 2015). Of that only 2.290 km^2 (1,3%) is used for horticulture. Of the horticultural area, 9.84 km^2 (0,0053%) is protected and under glass. Absolute and relative figures for other countries surely differ, but the example clearly shows that the plant side of aquaponics will only be able to substitute and thereby enhance a small fraction of our food production. Staple foods can theoretically be produced in CES under glass using hydroculture as demonstrated in NASA research (Mackowiak et al. 1989) and could surely also be cultivated in aquaponic systems, but due to the high investment needed for such production, it does not make sense to think of aquaponics replacing the production of these crops under the current global economic and resource conditions.

Hypothetical Modelling, Small-Scale Case Studies and Surveys Amongst Farmers

Early research on commercial aquaponics focused on evaluation and the development of specific, mostly research institute-led case studies. These first results were highly positive and optimistic about the future of commercial aquaponics. Bailey et al. (1997) concluded that, at least in the case of Virgin Islands, aquaponic farms can be profitable. Savidov and Brooks (2004) reported that the yields of cucumbers and tomatoes calculated on an annual basis exceeded the average values for commercial greenhouse production based on conventional hydroponics technology in Alberta. Adler et al. (2000) performed an economic analysis of a 20-year expected scenario of producing lettuce and rainbow trout and argued that the integration of the fish and plant production systems produces economic costs savings over either system alone. They concluded that an approx. $300.000 investment would have a 7.5-year payback period.

Technologically based dynamic optimization models are commonly used to represent production engineering relationships in aquaponics systems (Karimanzira et al. 2016; Körner et al. 2017). It is noticeable that so far hardly any different scales are considered, and previous studies like Tokunaga et al. (2015) and Bosma et al. (2017) are limited to small-scale aquaponics for the local production of food or are performed on data from research facilities, such as University of Virgin Islands' aquaponics systems (Bailey and Ferrarezi 2017). Furthermore, as Engle (2015) points out, the literature on the economics of aquaponics is sparse, with much of the early literature based primarily on model aquaponics. Without realistic farm data, such projections often are overly optimistic because they lack details on expenses beyond the obvious ones of fingerlings, feed and utilities and do not include the everyday risks involved in farming. In this research on the economics of aquaponics, production functions are only partially reproduced and questions of process-based optimization addressed only to some extent. Leyer and Hüttel (2017) demonstrated the potential for investment accounting as part of an initial analysis to capture various parameters of an aquaponics facility. Furthermore, Engle (2015) points to the difficulties of estimating annual costs to operate in aquaponics farms since many of these systems are quite new. She also points out that modelling is based on hypothetical situations and that more realistic farm data is needed, whereby the unexpected expenses are incurred daily, "from screens that clog, pumps that fail, or storms that cause damage".

As aquaponics started to grow both as a do-it-yourself (DIY) activity (Love et al. 2014) and as an industry (Love et al. 2015), research on real commercial farm case studies emerged. Specific case studies of aquaponics production were performed on commercial attempts, for example, in Puerto Rico (Bunyaviroch 2013) and Hawaii (Tokunaga et al. 2015), including also the case study of a small-scale aquaponics social enterprise (Laidlaw 2013).

With the continuous rise in the number of aquaponics growers, the first in-depth analyses of the state of the art of the industry emerged, focused primarily on the USA. These studies showed a less optimistic picture of the emerging industry. Love et al. (2015) performed an international survey amongst 257 participants, who in the last 12 months sold aquaponics-related food or non-food products and services. Only 37% of these participants could be named as solely commercial producers who gained their revenue from selling only fish or plants. Thirty-six percent of the respondents combined the sales of produce with aquaponics-related material or services: Sale of supplies and equipment, consulting fees for design or construction of aquaponics facilities and fees associated with workshops, classes, public speaking or agro-tourism. Finally, approximately one third (27%) were organisations that sold only aquaponics-related materials or services and no produce. The average aquaponics production site of 143 US-based producers was 0,01 ha. By comparing this to the overall hydroponic production in Florida (29,8 ha), Love et al. (2015) concluded that the size of aquaponics producers is significantly smaller than hydroponic production and is to a large extent still more of a hobby activity than successful commercial enterprises. In terms of water volume, the aquaponics farms reported comparable sizes as typical RAS aquaculture farms in the USA. Yet nearly a quarter of respondents (24%) did not harvest any fish in the last 12 months, and the estimated overall size of fish production was 86t of fish, which is less than 1% of the farmed tilapia industry in the USA.

According to the same study, aquaponics was the primary source of income for only 30% of respondents, and only 31% of respondents reported that their operation was profitable in the last 12 months. For example, the median respondent received only $1000 to $4999 in the last 12 months, and only 10% of respondents received more than $50,000 in the last 12 months. This led Love et al. (2015) to conclude that aquaponic farms were small-scale farms, which is comparable to agriculture in general, since farms with gross revenues of less than $ 50,000 made up approximately 75% of all farms in the USA and farms with less than $50,000 typically sold only around $7800 in local food sales – making it thus necessary to combine farming income with other sources of income. It is therefore not surprising that aquaponics, like small-scale farming, relies heavily on volunteer work. Typically, there were a large number of unpaid workers, family members and volunteers working on these small units, with an average of six unpaid workers per facility.

Similarly, Engle (2015) addresses the 2012 census where 71 aquaponics farms across the USA were reported which represented 2% of all aquaculture farms. Of these, only 11% had sales of $ 50,000 or more, compared to 60% of pond-based aquaculture operations that had sales of $ 50,000 or more. Additionally, Engle (2015) points out to the difficulties of obtaining data from these farms, for example, estimating annual costs to operate in aquaponics farms, since many of these systems are quite new.

In summary, from an economic point of view, there is a research gap in so far that there are no records and analysis available which include statements about economically viable systems. Further research is needed that would take into account (a) the production possibility curves (normative), (b) the combined analysis of fish and plants including feedback between both, (c) the economic efficiency in combination with optimising the business processes and feedback (simultaneous optimization production process and economic efficiency) and (d) the consideration of different scales (scale efficiency) against the background of the environmental sustainability of this agricultural system. In addition, there are no comprehensive and reliable data that combine key factors such as production volumes, factor entitlements and cost structures, scaling and sales strategies derived from existing real investments. Further profitability analyses should consider temporal aspects and risk whilst formulating normative benchmarks that in turn can serve as the basis for investment decisions.

Hypothetical Modelling Data from Europe

In Hawaii, Baker (2010) calculated the break-even price of aquaponics lettuce and tilapia production based on a hypothetical operation. The study estimates that the break-even price of lettuce is \$3.30/kg and tilapia is \$11.01/kg. Although his conclusion is that this break-even can potentially be economically viable for Hawaii, such break-even prices are much too high for most European contexts, especially when marketing through retailers and conventional distribution channels. In the Philippines, Bosma (2016) concluded that aquaponics can only be financially sustainable if the producers manage to secure high-end niche markets for fish and large markets for fresh organic vegetables.

Aquaponics on tropical islands (Virgin Islands and Hawaii) and warm, frost-free zones (Australia) contrasts highly to locations further away from the equator. The advantages in warm locations are the lower costs of heating and the seasonally even availability of daylight, thus allowing for potentially low-cost systems to economically survive. A frost-free location close to the equator with little to no seasonal differences makes it cheaper and easier to set up and operate a system year-round, which allows for semi-professional family business setups in those regions. Additionally, local production in these areas is valued higher since leafy green crops are either hard to store (e.g. Australia/heat) or difficult to transport to the customers (Islands) and generally have a much higher contribution margin than in locations such as Europe and Northern America.

Aquaponics can have several advantages in an urban context. Yet, advantages are only effective if the specific urban framework conditions are taken into account and if additional communication efforts are put in place. Peri-urban agroparks are presented by Smeets (2010) as a technically and economically viable solution for urban agriculture, offering synergy potential with existing industry through residual heat and suitable logistics as well as alternative inorganic and organic materials streams, for example, CO_2, from cement production. Rooftop aquaponics utilises "empty" spaces in urban areas (Orsini et al. 2017). Rooftops are often assumed to be free of cost "because they are there". Yet every space in the city is of high value. An owner of a building will always seek revenue for the space they offer, even the utilisation of vacant rooftops. A rooftop farm carries a high economic risk and changes may have to be made to the building (vents and logistics). Rooftops are also interesting for solar energy production with less risk to the operator.

Whilst aquaponics is often explicitly touted as a production technology suitable for urban environments and even areas with contaminated soil, the real estate cost is often completely underestimated. For example, official real estate prices in Germany can be examined via the online tool BORISplus (2018), revealing a significant gap between inner city limit prices and agricultural land prices. For example, peri-urban real estate within city limits in Dortmund, Germany, is in the 280 €/m^2–350 €/m^2 range, whereas agricultural land outside of the city limits is in the 2 €/m^2–6 €/m^2 range. In addition to that German building codes grant the privilege to farmers to erect agricultural buildings outside of the city limits. This legal and financial situation renders agricultural land in proximity to economic zones attractive for larger-scale aquaponic farms, leading to the above-mentioned concept of agroparks.

The placement of aquaponic farms raises challenges with customer perception. Citizens who have been interviewed about their preference of different urban agriculture concepts for inner city public land use showed a preference for usage that keeps the space accessible for citizens as well as a low acceptance levels for agroparks (Specht et al. 2016). The research results on the acceptance of aquaponics revealed a larger variance than the other potential utilisations, suggesting a citizen ambivalence due to a lack of information on the production method. Additional communication efforts are required as aquaponics is a highly complex and new production system unknown to most people in society including urban populations.

The potentials and risks of aquaponics in an urban context become clear from the paragraph above. Distinct strategies and contingency plans have to be developed in an urban context when planning to implement an aquaponics production facility.

Most of the data currently collected on commercial farmers is focused on locations outside of Europe. A sound economic assessment of aquaponics facilities in European latitudes and climates is difficult, because on the one hand only very few commercial plants exist in Europe and on the other hand technical equipment, scale and business models are very different in other parts of the globe, where commercial aquaponics is more widespread (Bosma et al. 2017). Whilst Goddek et al. (2015) and Thorarinsdottir (2015) provide a very good overview of European commercial plants and their challenges, they present only a few economic parameters such as (targeted) consumer prices, statements on "potentially" achievable income or break-even prices for production. Since these are only valid under the specific conditions of the investigated facilities, only limited statements can be transferred to other locations, even within Europe.

Whilst there are some specific assessments of productivity (e.g. Medina et al. 2015, Petrea et al. 2016), full market potential analysis and well-founded cost-effectiveness assessments are not known at the present time. In addition, there are initial studies on technical dynamic models using the methodology system dynamics such as Goddek et al. (2016) and Körner and Holst (2017). This illustrates how essential the availability of comprehensive data is in order to conduct sound profitability analysis.

One of the very few hypothetical modelling cases created with data from within Europe is Morgenstern's et al. (2017) model. They provided technical data from the pilot plant of the University of Applied Sciences of South Westphalia, which consisted of a commercial fish farm and a standard horticulture system. In this case, investment and full-cost calculations with comprehensive detailed technical data for systems in three different scales were modelled. Model calculations for operational costs for a start-up period of 6 years and investment costs as well as a simplified cost performance difference calculation have been conducted for three differently sized aquaponic farms rearing European catfish (*Silurus glanis*) and producing lettuce. The calculated sizes were derived from the pilot plant located at the University of Applied Sciences of South Westphalia and the aquaculture scale of the project partner. Modelled aquaculture sizes were 3 m^3, 10 m^3 and 300 m^3. A couple of general assumptions and simplifications were made for the calculations, which illustrate the above-presented critiques on the limitations of hypothetical modelling:

(a) Less than average production quality and production losses within the first 5 years have been considered. The profitability calculations are based on a matured and stable production process starting in year 6.
(b) Constant hydroculture production. The complete nutrient stream from the process water was calculated to be consumed by the hydroculture production of lettuce, regardless of seasonal differences and regardless of nutrient availability from the aquaculture.
(c) The hydroculture grow bed size has been calculated to be 60 m^2, 200 m^2 and 5.500 m^2.
(d) Heating demand for hydroculture and aquaculture has been approximated with a slightly modified methodology of KTBL (2009). The modelled location of the farm is Düsseldorf, Germany.
(e) Energy costs per kWh have been approximated for production with a combined heat and power (CHP) system with 15 ct/kWh (electricity) and 5,5 ct/kWh (heat), respectively. For simplicity, a CHP system has not been modelled.
(f) Direct marketing of the products was assumed. Fairly optimistic, but not overly optimistic, market prices have been calculated for the products. No extended marketing costs have been included in the calculation, since the marketing effort required to build a customer base and stable market have not been addressed in the project. Neglecting marketing costs assumes that market prices in direct marketing come at no costs and therefore constitute a major simplification of the calculation.
(g) No costs related to the real estate required for the farm have been included in the calculations. The rationale for this simplification is the vastly different costs for space depending on location and project context.

(h) Labour cost has been calculated at minimum wage, which is a strong assumption with regard to high levels of human capital required to run complex aquaponics systems.
(i) Mortality losses of 5% in the aquaculture system are compensated by overstocking at the start of each production cycle.

An analysis of the cost structure of the modelled production-sized aquaculture system shows that labour, fish feed and juveniles and energy are the main cost drivers, contributing roughly one third of the main costs each. At this point, it has to be emphasised that labour costs are calculated on a minimum wage basis and that costs for the occupied area of the farm have not been considered in the calculations (Fig.).

Electricity and heating costs offer potential for optimization. Pumps have a lifetime between 2 and 5 years. Inefficient pumps can be replaced with more efficient pumps in the natural machine life cycle. Cost efficiency gains for these kinds of optimizations are simple to calculate, and efficiency gains are also easy to monitor after implementation. Similar measures to reduce heating costs are relatively easy to calculate. For example, the costs and effects of additional insulation panels can be calculated, and also here the gains can be easily monitored.

Labour costs emerge as the main cost driver that shows significant optimization potential with upscaling. Larger-scale systems allow for the usage of labour-

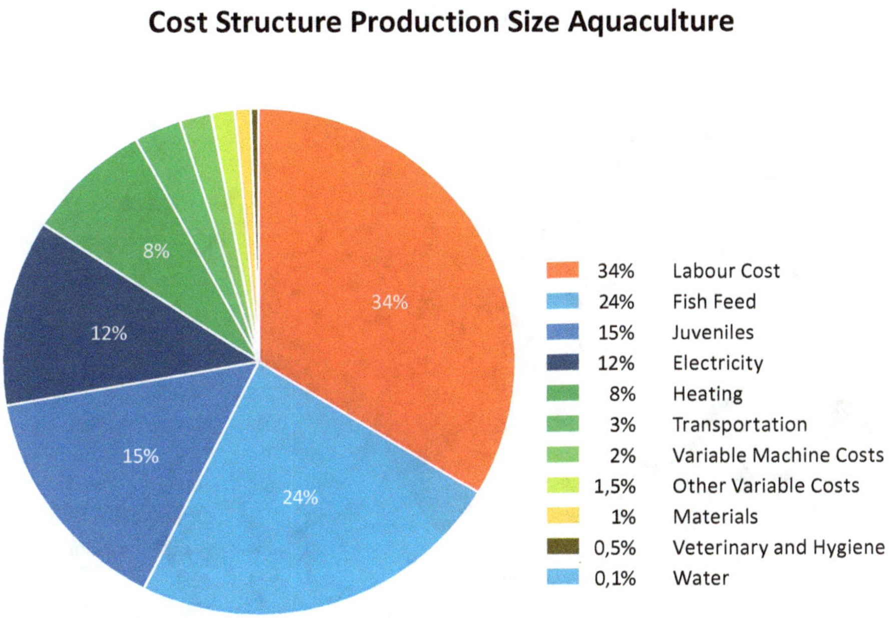

Fig. Cost structure for aquaculture side of an aquaponics system, hypothetical model from technical data from the pilot plant of the University of Applied Sciences of South Westphalia. (Based on Morgenstern et al. 2017)

saving devices, for example, automated graders or automated feeder-filling machinery. Profitability of these kinds of optimizations has to be calculated on a per-project basis.

Likewise, a cost analysis has been performed for the hydroculture part of the modelled systems. The main cost drivers are labour, seedlings and energy costs for lighting and heating. A higher operational maturity of the production, when the initial start-up learning curve has been mastered, can make room for in-house seedling production. The integration of this production step can offer cost optimization potential. Regarding cost reduction potential of the other cost drivers, energy and labour, the above-described situation is applicable for the hydroculture part as well (Fig.).

A cost performance difference analysis has been performed for the three system sizes, showing that the microsystem and the small system are not economically viable. There is no exploitable automation and rationalisation potential because of the extremely small aquaculture sizes and the small hydroculture sizes resulting in prohibitively high labour costs. Minimum quantity surcharge and transportation fees for the fish feed and similar effects for other cost categories put an additional financial burden on these two systems.

The production size system has a positive cost performance difference when real estate costs or tenures for the required land are not taken into account (Table).

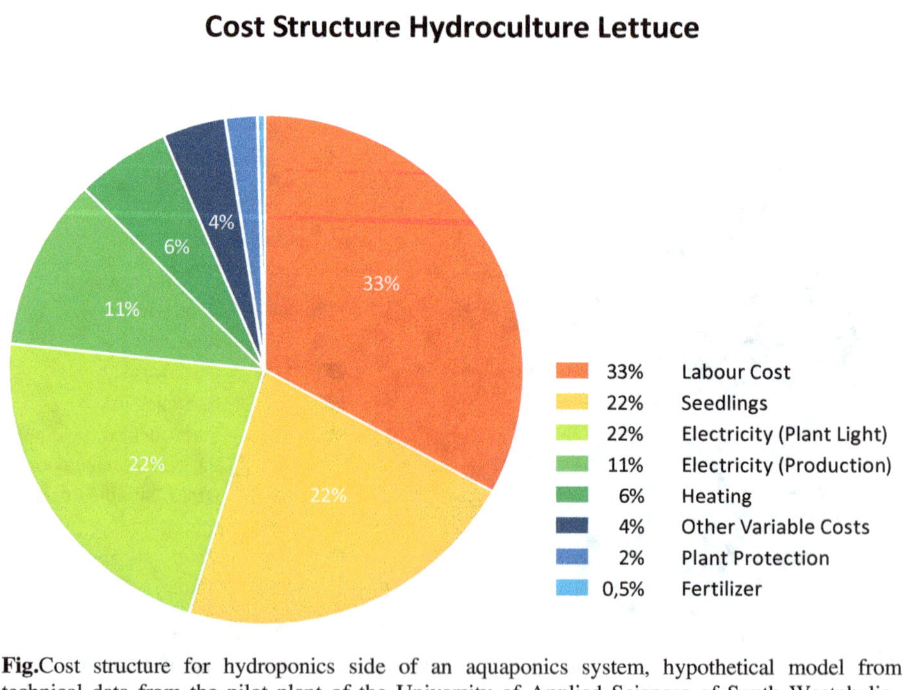

Fig. Cost structure for hydroponics side of an aquaponics system, hypothetical model from technical data from the pilot plant of the University of Applied Sciences of South Westphalia. (Based on Morgenstern et al. 2017)

Table Cost performance analysis of model calculation

Cost performance difference	Unit	Micro	Small	Production
Contribution margin aquaculture	€/a	−4173	−2566	114.862
Contribution margin hydroculture	€/a	691	13.827	541.087
Sum contribution margins		−3.483	11.260	655.948
Labour cost aquaculture	€/a	3.705	8.198	45.000
Labour cost hydroculture	€/a	3.148	8.395	179.443
Sum labour costs	€/a	6.853	16.593	224.443
Real estate costs tenure		n.a	n.a	n.a
Depreciation	€/a	7.573	15.229	185.269
Interest rate 2%	€/a	1.515	3.046	37.054
Cost performance difference	€/a	−19.424	−23.607	209.183

Source: Morgenstern et al. (2017)

Table Job creation potential

	Unit	Micro	Small	Production
Sum labour cost	€/a	6.853	16.593	224.443
Sum labour time	Days/a	46	111	1.496
Number of jobs		0,21	0,5	6,8

Source: Morgenstern et al. (2017)

The analysis additionally sheds light on the job creation potential of the respective systems. The model calculation was performed under the assumption that all the required overhead tasks of the enterprise arc handled by regular employees, an assumption that is rather optimistic with regard to the fact that the minimum wage has been used for the calculation.

One further assumption was made regarding the separation of jobs: The employees work on both parts of the sytem, the aquaculture and the hydroculture parts, in accordance to the work that is needed by the respective system. This requires an elevated skill set which puts another question mark behind the minimum wage calculation.

Even in the larger-sized production system, the number of jobs created is limited. The calculated number of jobs is congruent with experience from horticultural companies working with hydroponics, which usually employ between five and ten workers per hectare of greenhouse (Table).

Data on initial investments in aquaponics is on the one hand very difficult to come by and on the other hand even more difficult to compare. Some of the preliminary data collected from other sources on the initial investment needed to set up an aquaponics farm (see Table) below shows high differences between the initial investments in the systems, either real or in hypothetical modelling. Since the systems differ in the extreme amount of factors, it is highly problematic to draw any conclusions regarding the necessary initial investments. However, the initial investment in aquaponics does seem to be relatively high, which is reflective of the early stage of the industry. We estimate that an initial investment in a commercial aquaponics system in Europe starts

Table Estimated investment costs on aquaponics, various sources

Literature source	Total investment [approx. per m^2 of growth area]	Location	Aquaculture size and type	Hydroponic size and type
Bailey et al. (1997)	$ 22,642 [$ 226/ m^2]	Virgin Islands, USA	4 tanks Tilapia No heating	100 m^2 Lettuce DWC No greenhouse
Adler et al. (2000)	$ 244,720 [$ 240/ m^2]	Shepherdstown, WV, USA	19,000 l 239 m^2 Rainbow trout No heating ($ 122,80)	cca. 120 m^2 Lettuce NFT ($ 17,150) Polyethylene greenhouse with heating and lights ($ 78,770)
Tokunaga et al. (2015)	$ 217,078 [$ 190/ m^2]	Hawai'i, USA	75.71 m^3 Tilapia	1142 m^2 Lettuce DWC
Morgenstern et al. (2017)	€ 151.468 [€ 1067/ m^2]	Model location: Düsseldorf	3 m^3 European catfish	59 m^2 grow bed area 83 m^2 greenhouse Lettuce DWC
Morgenstern et al. (2017)	€ 304.570 [€ 650/m^2]	Model location: Düsseldorf	10 m^3 European catfish	195 m^2 grow bed area 274 m^2 greenhouse Lettuce DWC
Morgenstern et al. (2017)	€ 3.705.371 [€ 302/m^2]	Model location: Düsseldorf	300 m^3 European catfish	5.568 m^2 grow bed area 6.682 m^2 greenhouse Lettuce DWC

with at least 250 EUR / m^2 of growth area but can easily require a much higher investment, depending on the outside conditions, the system size and complexity and the length of the growth season aspired to (Table).

The experimental and pioneering status of commercial aquaponics is one reason why the financing of larger commercial-scale projects can be a challenge. Most aquaponic systems have been financed through research grants or through aquaponics enthusiasts. Personal communication with German banks that are traditionally strong in financing agricultural investments and that are therefore familiar with the intricacies of crop production and animal rearing revealed that they would not finance an aquaponics project due to lack of a proven and established business model (Morgenstern et al. 2017).

Aquaponic Farms in Europe

Thorarinsdottir (2015) identified ten pilot aquaponic units in Europe, approximately half of which were at the stage of setting up still rather small-scale systems for commercial production. Villarroel et al. (2016) estimated that the number of aquaponic commercial enterprises in Europe comprised approximately 20 companies. Currently, Villarroel (2017) identifies 52 research organisations (universities, vocational schools, research institutes) and 45 commercial enterprises in Europe. Only a handful of these, however, sell aquaponic produce and could be considered as an aquaponic farm. In 2016, as a spin-off from the COST FA1305 project, the Association of Commercial Aquaponics Companies (ACAC 2018) was founded, currently involving 25 companies from all over Europe, only about a third of which focuses on food production. Others offer mostly aquaponics-related services such as engineering and consulting (Thorarinsdottir 2015).

In the remainder of this chapter, we focus on the in-depth information from the few operating aquaponic systems in Europe. Within the time of the COST action (2014–2018), a number of organisations started and tested their first marketing attempts at commercial food production with the goal of making the sales of produce as their main income. We have gathered data with three in-depth qualitative methods: (a) site visits, (b) SME presentations at the COST FA 1305 conference (date, place) and (c) semi-structured in-depth interviews.

Site Visits

Within the COST FA1305 project, three European pilot aquaponic systems opened their doors for site visits of COST members:

- Ponika, Slovenia, Matej Leskovec (site visit made on 23 March 2016).
- UrbanFarmers AG, system in the Netherlands, Andreas Graber (site visit made on 6 September 2016).
- Tilamur, Spain, Mariano Vidal (site visit made on 20 April 2017).

Within the site visits, we asked questions about the type and size of the system, the initial investment, the types of fish and crops produced and the reasoning behind it and the selling experiences.

Presentations

COST FA1305 organised a special conference devoted to presentations of aquaponics SMEs in Europe at the International Conference on Aquaponics SMEs at the University of Murcia (ES) (19 April 2017). We only analysed the presentations and answers to questions from SMEs that presented their experiences with their own commercial aquaponics system:

- Bioaqua Farm, UK: Antonio Paladino.
- NerBreen, Spain: Fernando Sustaeta.
- Ponika, Slovenia: Maja Turnsek.

- Samraekt Laugarmyri, Iceland: Ragnheidur Thorarinsdottir.
- Tilamur, Spain, Mariano Vidal.
- Uit Je Eigen Stad, Netherlands: Ivo Haenen.

Follow-Up In-Depth Interviews

We made additional follow-up semi-structured in-depth interviews with the aquaponics SMEs, contacting also other European aquaponic producers, some of which declined to participate.

The final interviewers were:

- ECF, Germany: Christian Echternacht (interview made on 13 February 2018)
- NerBreen, Spain: Fernando Sustaeta (interview made on 7 April 2018)
- Ponika, Slovenia: Maja Turnsek (interview made on 30 January 2018)
- Samraekt Laugarmyri, Iceland: Ragnheidur Thorarinsdottir (interview made on 19 February 18)
- Uit Je Eigen Stad, Netherlands: Ivo Haenen (interview made on 8 February 2018)

The average interview lasted approx. 60 min. and included questions on the reasons for choosing aquaponics, type and size of the system, produce and fish type, experiences with growing, distribution and price, future plans and outlook for commercial aquaponics in the future. Below we will present the results together with the literature review on each of the areas of discussion (Table).

Horticulture Side of Commercial Aquaponics in Europe

Petrea et al. (2016) conducted a comparative cost-effective analysis on different aquaponics setups, utilising five different crops: baby leaf spinach, spinach, basil, mint and tarragon in deepwater culture and light expanded clay aggregate (LECA). Whilst the study was conducted in very small systems without taking into account any upscaling opportunity or potential, several aspects of the presented results are worth discussing. The grow beds have been illuminated in different lighting regimes with fluorescent bulbs and metal halide grow lights. The cost comparisons of the electricity illuminates the significant share plant lighting has on overall electricity cost. Furthermore, the analysis sheds light on the importance of sensible crop selection. Whilst tarragon is referenced as what is often called a "high-value crop" earlier in the text, the later economic crop yield analysis shows that other crops, basil and mint, generate a higher economic value per grow bed area (Petrea et al. 2016, p. 563).

Table General information on the interviewed aquaponics companies in Europe

Company	Location	Initial investment	Overall size of the system	Plant production size	Type of plants	Type of fish	Years of operation
ECF	Germany	1.3 mio EUR	1800 m^2	1000 m^2	Basil	Tilapia	2 years
NerBreen	Spain	2 mio EUR	6000 m^2	3000 m^2	Lettuce, strawberries, tomatoes, peppers	Tilapia	5 years the 500 m^2 pilot system, 1,5 years the large system
Ponika	Slovenia	100.000 EUR	400 m2 greenhouse + water reserve packaging and cooling area in a container	320 m^2	Fresh-cut basil, chives and mint	Carp, trout, big mouth bass (sold only carp and even that only as live fish)	2 years, after that stopped production
Samraekt Laugarmyri	Iceland	640.000 EUR		1000 m^2	TomatoLettuceHerbs	Tilapia, Arctic char	Commercial project is in planning phase
Uit je Eigen Stad	Netherlands	Embedded in larger project. Approx. 150.000 EUR–200.000 EUR	400 m^2	200 m^2	Head lettuce varieties	African catfishPink tilapia	Partially operational for 1 year. Never selling produce. Operation ceased

The comparison shows a wide span of economic value ranging from 5.70€/m²/cycle (baby leaf spinach) up to 2110€/m²/cycle (basil) and 23.00€/m²/cycle (mint). Realistically, these figures should not be taken as reference points for upscaled commercial productions, but they illustrate how different the economic output of the crop production can be. Table of the publication elaborates on the seasonal market price variability of the examined crops. The seasonal variations of these crops are rather moderate with slightly elevated prices in fall and winter months. Seasonal market price variations for produce and fruit with higher global market volumes like tomatoes and strawberries are usually much more pronounced, enticing producers to put effort into season extension at both ends. Artificial lighting for season extension is costly both investment-wise and regarding operational cost but might well be worthwhile, especially considering the inherent pressure to utilise process water from the aquaculture in the low light season.

No emphasis has been placed on product quality and marketability of the produced biomass within this study. Experience shows that the cultivation of certain crops is easier than the cultivation of others. Mint is generally regarded as an easy-to-grow crop, whilst the production of marketable basil is more challenging. Petrea et al. (2016) cultivated the crops in deepwater culture and ebb and flow grow beds with LECA substrate. The latter is particularly uncommon in commercial production, since it has close to zero potential for rationalisation and automation. Basil usually is produced in pot culture as opposed to a cut and come again production of mint which leaves the rootstock in the system with a faster regrowth of a marketable product. In addition to the different growth medium requirements, different crops are produced with differing temperatures, climate and light regimes. Optimal product quality can only be achieved with optimal cultivation techniques. It is important to remember that customers are used to premium quality and show little to zero tolerance for suboptimal products.

The horticulture side of commercial aquaponics faces high risks from infestations of diseases or parasites, which can be difficult to overcome because only biological controls can be used. Large risks are involved also since most aquaponic farms require a market that will pay higher-than-average prices for the crop. Finally, aquaponics seems to be very labour-intensive since even small scale aquaponics systems are complex because of their many compo-nents and requirements (Engle 2015).

For start-ups, it might be tempting to strive for the production of a wide spectrum of novelty varieties of plant species with fruity aromas or colourful leaves. Melon sage (*Salvia elegans*) or pineapple mint (*Mentha suaveolens* 'Variegata') are examples for these kinds of varieties. According to small-scale commercial producers in Soest, Germany (non-aquaponics; personal communication summer 2016), the market demand for novelty varieties has long been recognised by retailers and is being supplied for by their large-scale producers. This segment is not a profitable niche any more but rather a market that follows yearly trends. The switch by the Berlin-based aquaponics producer ECF from a wide spectrum of the aforementioned plant species in their start-up phase to a basil monocrop that is being marketed through the German retailer REWE reflects this situation.

Christian Echternacht from ECF, Berlin, reported in the interview about the difficulty to sustainably establish a local direct marketing channel for a wider range of products of limited quantity. Drawing from their first-hand experience, the company decided to shift the plant-side production to one crop, basil in pots, and to market this crop via one single retailer in over 250 supermarkets in the city of Berlin. Interestingly the regionally labelled product (Hauptstadtbasilikum/Capital City Basil) without an organic label is placed directly next to organically labelled basil from non-regional sources and is reported to generate higher sales despite slightly higher prices.

NerBreen based in Spain with its 6000 m^2 size is currently the largest system in Europe. It is more focused on the aquaculture element and includes aquaponics as one of several means of water filtration, but the plant production is still 3000 m^2 and produces enough to create a market. They are currently undergoing the second season of production within the farm, having 5 years previous experiences with a smaller pilot plant (overall size of the pilot farm, 500 m^2). In the winter season, they now grow fresh garlic, strawberry plants without fruit (since the plants need to be maintained for 3 years) and four different types of lettuce. In the summer, they replace the garlic with cherry tomatoes and peppers but keep the strawberries and the same lettuce varieties. Since this is only the second season, it is very difficult for them to provide an average amount of produce. The last winter was very cold, and it significantly affected the growth of the lettuce. In the first season, when they were still trying to improve and gain experience, they produced about 3t of strawberries, 5t of tomatoes and 60,000 lettuce heads. Their hopes are to ramp up production, whilst at the same time, their strategy shifted from putting a focus on the quantity to quality and variety instead. In the first year of their production, they had a good season with tomatoes in terms of quantity – but the overall market was flooded with tomatoes, and the price was subsequently too low. They adjusted to this problem with a focus on more selected, niche varieties of cherry tomatoes since the price is better, and they do not to want to compete with quantity but with quality, thus trying to reach a higher price with the retailers.

Locally produced crops and niche cultivars seem to be the main directions for crop selection in the European countries. The Slovenian company Ponika set out to sell fresh-cut herbs in their 400m^2-sized system since providing niche products in the Slovenian market with no other local producer of fresh-cut herbs. The company based their rationale on three main reasons. The first was that the data available, albeit scarce, from US aquaponics farms showed that fresh-cut herbs seemed to be the crops that succeed well in aquaponics and gained a higher price in the market. The second was the years of positive experience within the small-scale DIY aquaponics garden with these crops. In addition, the third was the positive feedback received from fresh-cut herbs distributors in Slovenia regarding the interest in the crops. The company set out to produce fresh-cut herbs and managed to sell them to Slovenian gastro-distributors for two seasons, narrowing the initial number of crops from six to three: fresh-cut chives, basil and mint. Other fresh-cut herbs they tested proved to be either too sensitive, or there was too small and infrequent demand on the market. The plan was to first sell the fresh-cut herbs to the gastro-distributors and then gradually proceed to selling these to large-scale retail chains. The reality, however, showed significantly large risks in the production (e.g. powdery mildew with basil and yellow tips with chives) and too small a system to be able to secure a steady uninterrupted production as requested by the large-scale distribution chains. Although the margins would be higher, Ponika never started to sell to the retail chains since the contracts with large retail chains included financial penalties in the case where the farm could not deliver the orders. Additionally, the retail chains made weekly orders thus not allowing for appropriate planning, and in some cases, the excess that was not sold had to be collected by the farmer, and this discarded produce was expected to be deducted from the overall order even though the over-ordering was on the side of the retailer.

The main reasons for the Slovenian-based company Ponika to stop their operation were the combination of high risks accruing from the labour needed to cut, screen and package the produce combined with the average gained price of 8 EUR / kg of fresh-cut herbs (packaged in 100 g bundles) not providing enough of an economic return to cover the extra workload. Since the company was the only Slovenian company in the Slovenian market, the gastro-distributers were willing to take their produce over the imported produce – yet only if the prices were equal to what the prices of international competitors were on the market. With a high percentage of fresh-cut herbs sold in the European market being delivered from North Africa, the high extent of labour costs for aquaponically produced fresh-cut herbs meant that a small-scale system could not compete with the prices as set by the extensive fresh-cut herbs farms in warmer regions with lower labour costs, even when including transportation costs. This shows that even when there is a niche in a local market, there are often specific reasons for local producers not filling the niche. In the case of fresh-cut herbs in Slovenia, this was the high cost of labour for too small a niche market.

Aquaculture Side of Commercial Aquaponics in Europe

Starting a business in temperate climate regions of Europe or Northern America requires a larger investment since the systems have to be kept frost-free requiring more electrical energy for plant lighting when operated throughout the year. In Europe, there are two strong horticultural production powerhouses, one in Westland/NL and the other in Almeria, southern Spain. The market concentration is high and contribution margins are slim. As a result, some aquaponic producers presumed that in aquaponics the contribution margin from aquaculture is more interesting than that of horticulture, which is probably why some of the few commercial operators chose to oversize the aquaculture part of the setup. This can lead to technical issues because a larger quantity of nutrients than required by the plant side is being produced in the aquaculture side. The excess process water has to be discarded

(Excursion Graber 2016 and Interview Echternacht 2018), putting the sustainability claims of aquaponics in question. Christian Echternacht from ECF reports that the contribution margin of the aquaculture has been overestimated in early calculations, rendering the oversizing of the aquaculture part of the farm counterproductive for the overall profitability of the farm.

Numerous different fish species have been reported to be produced in commercial aquaponics in Europe. Popular species for aquaponics production are tilapia, African catfish, largemouth bass, jade perch, carp and trout. There is no known commercial aquaponics farm currently rearing European catfish, but researchers at the University of Applied Sciences of South Westphalia (Morgenstern et al. 2017) found this species to be suitable for aquaponics production. The selection of fish species is influenced by a large number of different project-specific parameters. Most importantly of course are the market needs, price and distribution options. Within Europe, coastal regions have a traditionally strong market for marine fish with a diverse set of species and products. This creates a marketing challenge for freshwater aquaculture production. Ivo Haenen from *Uit je Eigen Stad*, Rotterdam, and Ragnheidur Thorarinsdottir from *Samraekt Laugarmyri*, Iceland, talked about this effect in their interviews. Rotterdam customers are used to a rich and diverse supply of marine fish products, making it difficult to market freshwater tilapia and African catfish. The marine wild catch tradition is so ingrained into Icelandic culture that the aquaculture aspect of aquaponics is probably not going to be actively promoted in future aquaponics projects.

Tilapia, one of the most commonly used fish species in aquaponics in the USA (Love et al. 2015), is a fish species that is not commonly known in Europe. As the experiences from NerBreen in Spain show, European tilapia aquaponics producers face a double marketing challenge: Their marketing attention needs to be put not only towards building customer awareness on the benefits of aquaponics production but also on the benefits of this relatively unknown fish species.

The suitability of the selected species for elevated water temperatures is another important factor. Fish are poikilothermic; thus their growth and consequently their production yields speed up with higher water temperatures. But elevated water temperatures require more energy, which, depending on the selected energy source for heating the process water, is connected with higher operational costs. Therefore, the positive effect of higher yields has to be balanced with the elevated costs for heating the water. From this perspective, it is desirable to tap into the potential of residual heat usage from adjacently located power plants or industries. These locations, however attractive and sensible they may be from an economical and ecological point of view, might pose a challenge for the overall marketing of the farm and its products. Industry sites are usually not idyllic and emotionally attractive, and worse still in case of anaerobic sewage plants or similar industries, they might even appear to be repulsive. Consequently, the available locations, and the context the farm can reasonably be placed in, is one factor for species selection.

The influence the different fish species have on plant yield and quality has not yet been completely researched. Knaus and Palm (2017) conducted experiments comparing plant yield in two identical aquaponics systems with identical operating parameters rearing tilapia and carp and found that plant performance with tilapia was better than with carp. These results show that there is indeed a difference in fish-plant interaction, but these have not been researched for a wider range of different species. In addition, the potential for fish polyculture, where two or more different fish species are reared in the same aquaculture cycle, has not yet been systematically researched.

One of the important operational factors for the fish selection is juvenile availability. Most of the commercial aquaponics producers buy juveniles from hatcheries. One notable exception is the company Aqua4C in Belgium that produces jade perch juveniles and uses these fish in their aquaponics system. A common recommendation is to select a species with at least two known suppliers with significantly larger capacity than the projected demand for the aquaponics farm. The rationale behind this recommendation is risk mitigation. If the supplier of juveniles experiences production issues and cannot deliver, the whole aquaponics production is in jeopardy.

As with the horticulture side of aquaponics, similarly the aquaculture part of aquaponics faces high technical risks, such as the death of the fish due to electricity outages, as reported by both Ponika from Slovenia and NerBreen in Spain. Ivo Haenen, former operator of the aquaponics system of the Urban Farm "Uit je Eigen Stad" in Rotterdam, reports that the heating system of the initial system setup was not dimensioned appropriately. An unexpected period of cold weather led to lower than tolerable process water temperatures resulting in losses in the aquaculture part of the operation. These kinds of instances have to be attributed to the pioneering character of early commercial aquaponics operations in Europe. The presented cases illustrate why Lohrberg and his team classified aquaponics in the "experimental" category of the seven identified business models of urban agriculture (Lohrberg et al. 2016).

Public Acceptance and Market Acceptance

The future of aquaponics production depends on public perception and the associated social acceptance in important stakeholder groups (Pakseresht et al. 2017). In addition to potential aquaponics plant operators, players at the wholesale and retail level as well as gastro-distributors and collective catering are important actors in supply chains. Moreover, consumers are key actors as they bring in the money into the supply chain at its end. Even though they have no direct economic stakes in aquaponics production, the general public as well as political and administrative bodies are important aspects to consider. The necessity of involving the aforementioned stakeholders is based on studies such as Vogt et al. (2016), who show that suitable framework conditions are an important basis for the establishment of innovative processes in food value chains. Technical developments without involving stakeholders run the risk of non-acceptance at the end of the research and development pipeline. In general, they build on a comprehensive understanding of a marketing philosophy with a multi-stakeholder approach.

For aquaponics, there is still no knowledge about the conditions that promote the dissemination of this technology. Although the technology used in aquaponics installations for freshwater fish farming in tanks is also used in aquaculture, until now this is unknown to a large part of society (Miličić et al. 2017). With regard to consuming plants from aquaponics, there is scepticism regarding their contact with fish water (Miličić et al. 2017). Preliminary studies based on a small sample regarding the acceptance of aquaponics products by potential consumers indicate that the requirements for products from aquaponics facilities go far beyond what the previous purchasing behaviour of fish products suggests (Schröter et al. 2017c). Based on the results of Schröter et al. (2017a), first hints on the effect of information on the acceptance process are available. These need to be further explored by means of perceptual and impact analyses of various information and presentation variants (e.g. textual facts, images, word-image content) and validation on the basis of representative samples. In addition, previous research has focused on citizens in general and on potential consumers. Studies on the acceptance of other important stakeholders such as potential plant operators, food retailers and public catering as well as political and regulatory actors and the general public are lacking completely.

First analyses of the consumers' response on aquaponics indicate that consumers showed a positive attitude towards aquaponics, with food safety issues being the major consumer concern in Canada (Savidov and Brooks 2004). Initial preliminary work on the willingness to pay for fish products from aquaponics was carried out by Mergenthaler and Lorleberg (2016) in Germany and Schröter et al. (2017a, b) on the basis of non-representative samples in Germany. Part of these studies show a relatively high willingness to pay for fish products from aquaponics. However, these results are based on small samples and cannot be generalised because the willingness to pay has been compiled from a specialist target group (see Mergenthaler and Lorleberg 2016) or in connection with the visit to a greenhouse for tropical and subtropical plants grown using aquaponics (Schröter et al. 2017a, b).

According to Tamin et al. (2015), aquaponics products are green products. A product is defined as *green* when it includes significant improvements in relation to the environment compared to a conventional product in terms of the production process, consumption and disposal (Peattie 1992). Based on the "theory of planned behaviour (TPB)", consumer acceptance of aquaponics products as innovative green products has been examined by Tamin et al. (2015) with closed-ended questionnaires in Malaysia. From a set of different behaviour-influencing factors (relative advantage, compatibility, subjective norm, perceived knowledge, self-efficacy and trust), two factors have been identified as having a significant impact: Relative advantage and perceived knowledge. The *relative advantage* describes how far buying behaviour is influenced by superior product qualities compared to conventional products. The aquaponics products were perceived fresh and healthy, and this perception led to a buying advantage. The *perceived knowledge* relates to how much the customer knows about the production method. The more the customers were familiar with the method, the more likely they were willing to buy aquaponics products. There was no correlation in the category subjective norm, which relates to how much the buying decision is influenced by the opinion of friends and family. Interestingly there was no correlation for the factor of *compatibility*. This factor relates to how much the product buying experience is compatible with the customer lifestyle. It seems as if the product to market process in Malaysia is not very different for aquaponics products and conventional products. So while it is questionable whether the results of this study can be safely transferred to European markets, a base message is that education about the production method and communicating the beneficial effects regarding the freshness of the food and the benefits for the environment are important marketing activities (Tamin et al. 2015).

Zugravu et al. (2016) surveyed the purchase of aquaponics products in Romania. Customers were influenced by friends and family. This dimension, subjective norm, showed no correlation in the Malaysian survey. The survey finds that consumers have a general good overall image of aquaponics. They think that the products are good for their health and that they are fresh. The paper describes a discrepancy between the perception of fish from aquaculture and wild catch and the perception of aquaculture. Retailers think that farmed fish can have a negative image, but actually aquaculture itself does not really have a pronounced image. The lack thereof is perceived by retailers as a marketing risk, yet it is described in the paper as giving the potential for positive branding through targeted communication. As a recommendation, the paper concludes that the retailers should build on the trust that the consumers showed when purchasing these fish products and should label the aquaculture fish as "healthy and fresh" (Zugravu et al. 2016).

Interestingly both Tamin and Zugravu had a significant higher questionnaire return count from women (Tamin et al. 2015; Zugravu et al. 2016). This raises the question about gender differences in aquaponics marketing. Although quantitative studies include gender as an independent variable in their analyses, no systematic and consistent patterns have been found yet. This asks for more research explicitly addressing gender aspects.

According to Echternacht from ECF in Germany, whose main business model is to set up aquaponics systems, marketing is the component that is usually most underestimated by their potential clients. ECF Farmsystems builds on this experience and surveys their potential customers for their intended marketing and distribution goals. If they have an existing business with actual production and established marketing channels, then the customer is very interesting. Idealistic customers who think that the products are going to market themselves are treated with caution.

Depending on the intended target group, different scales of production units might be favourable. Whilst some consumer segments prefer small-scale production possibly linked to short transportation distances and local production, there might be other consumer segments more interested in resource efficiency and low-cost production which can be realised in rather large-scale production units linked to waste energy and waste heat sources. Results from Rostock show (Palm et al. 2018) that small-scale systems with simple technology can make sense. Medium-scale systems require all the maintenance and the operational expenses of larger-scale systems, but do not have the benefit and output of large-scale systems. Conclusions from their experience show that you should either go small and achieve high prices in local markets or go to larger-scale systems with respective exploitation of economies of scale allowing for price reductions. Bioaqua from the UK is one of the rare European aquaponics companies that decided to follow the path of small-scale production with simpler and cheaper systems and providing the added value via catering and finding niche products for direct distribution to restaurants.

There might be other consumer segments displaying high preferences for fish welfare who therefore have to be targeted with fish from production units that conform to these ideals. As Miličić et al. (2017) show, consumers can express unexpected aversions, such as vegans expressing highly negative attitudes towards aquaponics. As pointed out in the literature, some facets of aquaponics may arouse high emotional involvement, such as the aesthetics of the aquaponics system (Pollard et al. 2017), level of mechanisation (Specht et al. 2016), soilless crop production (Specht und Sanyé-Mengual 2017), fish welfare (Korn et al. 2014), concerns about health risks due to the water recirculation system (Specht und Sanyé-Mengual 2017), or negative emotions bordering on disgust, because fish excrement is used as fertiliser for vegetables (Miličić et al. 2017). In this context, the perception and evaluation of aquaponics and its products may be based on unconscious processes rather than on careful consideration of logical arguments.

For some consumer segments, plants from aquaponics are innovative and interesting, and for others the link between fish and plant production might not be acceptable. This is also shown by ECF in Berlin: ECF decided to modify their initial production and marketing strategy. In the beginning, they attempted to produce a wide range of crops and market them directly on location. Nevertheless, according to Christian Echternacht (Interview Feb 2018) the marketing effort is simply too large. From their experience, the customers do not want to visit too many locations with only a few products at each location. Therefore, ECF decided to produce only one crop, basil, that is being marketed through a supermarket chain. Their experience as well as more comprehensive literature reviews shows that depending on the degree of meeting customer expectations, different levels of willingness to pay can be achieved and therefore achievable market prices are highly context specific.

Similarly, Slovenian-based company Ponika first attempted direct distribution of their fresh-cut herbs to restaurants in Ljubljana. But, just as with individual customers, restaurants were also averse to direct ordering even if the price was lower. For the restaurant managers, the time and effort needed to order individual products was much too high a price to pay, and they were not willing to order directly. They preferred to stay within their own gastro-distributors, whereby they could make their overall purchase in just one order.

The experiences of ECF from Berlin and Ponika from Slovenia described above are in line with previous experiences in the marketing of organic food products. To sell these products locally through direct distribution will only be possible for a small part of the products. Even though many consumers want to buy local and/or organic products, they often want to make their purchases as conveniently as possible. This means that shopping has to be efficient in order to fit into their daily schedule. As shown by Hjelmar (2011) for organic products, the availability of these food products is important for consumers because most of the consumers are pragmatic. They do not want to go to several stores in order to get what they want. They want to buy their products conveniently in a nearby supermarket and if the supermarket does not have a wide selection of organic products, many consumers end up by buying conventional products (Chryssohoidis and Krystallis 2005). Similar experiences can be described for consumers buying regional products in Germany (Schuetz et al. 2018). The same will presumably apply to aquaponics-grown products. If these products will not be available in supermarkets, aquaponics will probably remain niche production.

Organic food shoppers constitute a special potential target group of aquaponics. Indoor production of vegetables might require less or no pesticide applications, but soilless cultivation of plants is not an option in today's legislation on organic agriculture (cf. Chapter of this book). Therefore, aquaponics in its strictest sense will not provide the necessary characteristics to be eligible for certification as organic production and no organic labels would be allowed on aquaponics products. Therefore, either policymakers have to be lobbied to induce changes in organic legislation, or organic shoppers have to be educated in this rather complicated issue. This aspect is also important in the regard that organic classified products usually achieve higher market prices than conventional products, and such certification would make the aquaponics systems more economically viable. If aquaponics-grown products can be sold at the same prices as organic products, under certain conditions, the payback period of aquaponics systems can be reduced by less than half (Quagrainie et al. 2018).

Besides product marketing, services surrounding aquaponics production can generate additional income streams. The high level of innovativeness of aquaponics generates high levels of interest, which can be exploited in different service offers which included paid-for aquaponics visits, workshops and consultancy services around the establishment of new aquaponics systems. There are several examples of aquaponics facilities venturing in this direction:

- ECF provides business consultancy for the establishment of new aquaponics systems.
- UrbanFarmers, Den Hague, offered paid visits to the facility as well as an event location. (Note: The project has now ceased).

Besides adjusting production systems to customer expectations in a comprehensive marketing concept, communication strategies also play a role. Up to now, knowledge about aquaponics in the society is weak (Miličić et al. 2017; Pollard et al. 2017). When acquiring information, different variants of information and of information representation will significantly influence the public perception of this innovative technology. To satisfy the stakeholders' information demands, different channels of communication and different information materials can be used.

Diversification strategies are required that include workshops, visitor guides and other services. There are opportunities for new and alternative business ideas. Examples of innovative communication approaches by some commercial aquaponics operators show the specific challenges associated with aquaponics:

1. ECF, Berlin: Choosing red variety of tilapia. Branding as "Rosébarsch" at the beginning of sales. Inspired by a customer's branding in a restaurant, ECF rebranded to "Hauptstadtbarsch" (capital city perch) in the meantime (Interview Echternacht 2018). Thereby regional branding is put in the foreground of communication rather than the inherent product quality oriented at the colour of the fish meat.
2. Aqua4C, Belgium: They introduced jade perch from Oceania into the European market. Aqua4C developed a branding as "Omega Baars" thus taking a novel food approach. They implicitly market the regionally unknown fish species as healthy while carefully avoiding to make any health claims.
3. Ponika, Slovenia: Marketing of produce from aquaponics has been difficult. The situation was complicated as the aquaponics farm was located far away from the market which was too far for a quick visit and also no other attraction was nearby. They concluded that without the possibility of visits, it would be difficult to secure the farm additional revenue sources or marketing directly to consumers. In their marketing approach, they thus first targeted gastro-distributors with a focus on quality and local production for a competitive price. They did not dare to target individual consumers via supermarket chains due to having too small of a system and subsequent inability to secure a steady, large enough volumes of production. Thus they sold fresh-cut herbs directly to gastro-distributors, whereby the price and local production played the most important role. Their experience showed that gastro-distributors liked the story of innovative food production and they liked helping young people in their start-up business. So they were supportive in the sense that they adopted their purchasing process by taking up the produce when it was available and ordering from foreign sellers when it was not. In general, however, they were not very interested in the sustainability character of aquaponics – in other words, they did not care how the fresh-cut herbs were produced but rather that they were locally produced and had appealing packaging (1 kg and 1/2 kg) where the local character of the production was emphasised. Thus in their experience with retailers, a company story of young innovators worked the best. Customers at the retail level furthermore did not like the connection with hydroponics as they mixed aquaponics and hydroponics. In Slovenia, the customers are wary of hydroponics, and the Ponika company needed to tackle the challenge of changing the consumer's perception from hydroponics, which has a negative image as being "unnatural", into aquaponics and create a positive image of aquaponics. Additionally, the selection of fresh-cut herbs for the individual consumers proved to be problematic, since the health benefits were not important enough in fresh-cut herbs as people just do not eat that much of those to care enough about, for example, pesticide-free production.

4. NerBreen, Spain: NerBreen is focused more on the aquaculture element of their business, since 70% of their business model represents the revenue from selling the fish. Yet they provide extensive marketing of both the fish and the vegetables. In both cases, they try to target the individual consumers via retail chains, preferably those retailers that target consumers who are willing to pay the premium price for higher local quality, aiming for approximately 20% higher prices than average. They face challenges in the marketing of both vegetables and tilapia. With vegetables they focused on fresh garlic and cherry tomatoes because they could reach higher prices due to smaller competition in those areas. Their marketing efforts include well-designed packaging with leaflets, whereby they explain the sustainability benefits of aquaponics. Here, they focus both on securing premium quality and adding additional story to their branding. When selling tilapia, they face a bigger challenge. The Spanish consumers currently have a negative perception of tilapia since they either mistake it with pangasius which is considered as a cheap and low-quality fish or they think it is imported from intensive aquaculture from the Far East and similarly ancd supposedly to be lower in quality. Within their marketing efforts, NerBreen thus needed to change this negative image and is focusing on providing information on the fact that this tilapia is locally produced, whereby both the water quality and the fish feed quality are of highest consideration, resulting in a high-quality fish product.
5. Urban Farmers, Netherlands: In an effort to generate additional income streams, they established visits of aquaponics production facilities. It is, however, currently questionable if the visitor business is economically sustainable. Questions arise whether the visitor stream will wane when the hype around aquaponics settles or once "everyone" has already seen it. Apart from visitors, other income streams are already tapped: Rooftop Farms offer gardening workshops. *(It should be noted that Urban Farmers in the Hague ceased to trade.)*

Print media and social media are suitable for public education, as well as thematic workshops, guided farm visits and tastings of aquaponics products (Miličić et al. 2017). However, information provision will be successful only if it meets the information needs of the target audience. Stakeholders, such as representatives of national governments, different associations (e.g. organic farming associations), plant operators or plant manufacturers are probably more interested in comprehensive factual information. For citizens' and consumers' information, focusing on emotion and entertainment could be more attractive. With regard to this target audience, pictures combined with concise text messages are particularly suitable for information transfer. For these stakeholders beyond conscious information perception and information processing, also unconscious effects play an important role. Different frames, that mean different presentation formats of the same information, can influence the recipient's behaviour in different ways (Levin et al. 1998). For a better understanding of the unconscious processes that may influence the stakeholders' behaviour, neuroeconomic research methods in association with traditional methods of market research are useful tools. Eye tracking makes it possible to answer questions regarding visual perception in an objective way. Combined with other empirical methods of communication research, especially qualitative and quantitative surveys, it is possible to conduct complex perception and impact analyses. As a pilot study by Schröter and Mergenthaler (2018a, b) shows, attitudes towards different aquaponics systems are related to the gaze behaviour of the study participants whilst viewing information material about aquaponics.

This underlines the importance of a careful and target group-oriented design of information material about aquaponics. The possible solutions are that either production planning has to accommodate and to add the additional revenue sources or direct marketing by growing a large variety of different crops, thereby further complicating the production process. Yet as, for example, ECF in Germany shows, they started with a variety of vegetables but decided to focus only on basil and sell it through one large retail chain. Another possibility is to build strategic alliances with other regional producers in order to achieve innovative marketing and distribution strategies. In general, however, we can conclude that the marketing aspect of commercial aquaponics is one of its most important challenges and one in which European aquaponics farms had to undergo a number of changes in attempts to try and find the right product-market fit. It remains to be seen, however, whether this product-market fit has been found and how stable it will remain.

Conclusion and Outlook

As discussed in this chapter, economic evaluations of aquaponic systems are still a very complex and difficult task at present. Although aquaponics is sometimes presented as an economically superior method of food production, there is no evidence for such generalised statements. Up to now, there is hardly any reliable data available for a comprehensive economic evaluation of aquaponics. That is partly because there is not "one aquaponics system", but there exist a variety of different systems operating in different locations under different conditions. For example, factors such as climatic conditions, which mainly affect the energy consumption of the systems, wage levels, workload required for operating the systems, and legal conditions have to be considered on the cost side. On the revenue side, factors such as the chosen fish-plant combination with its specific product prices, the option to manage the systems as organic production as well as the long-term public acceptance of the aquaponics systems and their products have an impact on the economic assessment. Not least, the economic evaluation of aquaponics in its strictest sense should be done in comparison to recirculating aquaculture systems and hydroponics systems as stand-alone systems.

Aquaponics constitutes a major communication challenge as a rather unknown food production system with high innovation levels and in most cases with high technological inputs. As food consumption in advanced societies is increasingly linked to some form of naturalness, major challenges in the communication of aquaponics systems and products can be expected. The limited evidence available suggests that this challenge can be managed under certain framework conditions but this needs high time, as well as financial and creative inputs. It has to be acknowledged that the reported high prices for aquaponics products only come at considerable costs of brand establishment. As any economic viability of aquaponic systems will critically depend on achievable prices, more research is needed to understand the different determinants of the customers' willingness to pay for aquaponics products.

Location decisions for aquaponics farming are a key determinant of economic viability as many production factors related to aquaponics production are not flexible in terms of space. This relates particularly to land. Aquaponics as a land-efficient production system can only count on this advantage in land-scarce regions. Comparatively, rural areas with relatively low land prices therefore cannot generate sufficient incentives unless there are other site-specific advantages, for example, waste energy supply from biogas plants. Though being land-efficient in general, aquaponics in urban contexts still competes for highly limited land resources. In functional markets, land would be allocated to those activities with the highest profits per unit of land and it is highly questionable if aquaponics will be able to compete with very efficient industrial or service-oriented activities in urban contexts. Therefore, aquaponics seems to fit only in urban areas that provide aquaponics with a competitive advantage over competing potential activities.

Extending the definition of aquaponics and including aquaponics farming as introduced by Palm et al. (2018) might align aquaponics much closer to traditional economic farming analyses. This wider definition of aquaponics refers to process water that is used for fertilisation combined with irrigation on fields. With this wider interpretation of aquaponics, it becomes possible to produce staple foods in aquaponic production systems. Since the nutrient absorption capacity of the agricultural area might be limited in some regions, this definition implicitly positions aquaponics as a competitor to pig, beef and poultry production. Since aquaculture uses less resources than pigs, beef and poultry with respect to the final output, this could become a viable option.

In traditional economic farm analyses, there is a strong technological and conceptual separation of animal and plant production. With less technological interaction in aquaponics farming as compared to aquaponics in its stricter senses, there will also be less complex economic evaluation as the two systems of fish and plant production can be modelled separately. To link the systems economically, system internal prices would have to be determined, e.g. prices for nutrients brought from fish production to fields of plant production.

Another issue is the prices obtained for final products from aquaponics farming. Evidence on achievable prices from such kind of production systems are completely lacking thus restricting reliable estimations on the economic viability as yet. With a stronger separation of fish and plant production, it might be feasible to use prices for conventional aquaculture products and conventional prices from plant production. This would assume that there is no price premium for aquaponics farming. To test whether this really is the case, price experiments combined with different communication tools have to be implemented.

From a communications perspective, there is the question of the perception of aquaponics farming as being superior to traditional farming approaches. At first glance, aquaponics farming might look like conventional livestock keeping only using a different type of animal. Communication efforts will have to focus on the higher efficiency levels of aquaculture as compared to other types of livestock production. Advertising plant products from aquaponics farming as being superior to products from conventional plant production might be a challenge and requires further in-depth analyses. However, one advantage with regard to communication might be the fact that a stronger separation of fish and plant production in aquaponics farming could facilitate organic certification. Organic labelling is expected to be a further advantage in communication efforts related to aquaponics. (It should be note that in the UK, at least, organic certification is tied to growing produce in soil and thus a different and special type of certification may need to be identified. For more on the issue of aquaponics and organic certification).

Finally, it is important to note that the interviewed European aquaponic companies, even those who abandoned their commercial aquaponics farming, remain hopeful for the future of aquaponics. They opted for aquaponics because of its sustainability potential and they still see that potential. They, acknowledge, however, that adoption of aquaponics is a gradual and long-term process, which cannot be just simply repeated in different locations but should be adapted to the local environment. As such, aquaponics remains one of the potentially sustainable technologies of the future, one which cannot (yet) be said to be able to properly compete on the market with its competitors, but one that will continue to need more public support and one whose adoption is determined not only by its commercial advantages, but much more on the public determination and goodwill. As stated in Chapter of this publication, asking the question "under what circumstances can aquaponics outcompete traditional large scale food production methods?" is not the same as asking "to what extent can aquaponics meet the sustainability and food security demands of our age".

References

ACAC Association of Commercial Aquaponics Companies (2018) 10 Dec 2018. https://euaquaponicshub.wordpress.com/association-of-commercial-aquaponics-companies-acac/

Adler PR, Harper JK, Wade EM (2000) Economic analysis of an aquaponic system for the integrated production of rainbow trout and plants. Int J Recirculating Aquac 1(15):15–34

Baker AAC (2010) Preliminary development and evaluation of an aquaponic system for the American Insular Pacific. [Honolulu]:[University of Hawaii at Manoa], [December 2010]

Bailey DS, Ferrarezi RS (2017) Valuation of vegetable crops produced in the UVI Commercial Aquaponic System. Aquacult Rep 7:77–82

Bailey DS, Rakocy JE, Cole WM, Shultz KA, St Croix U (1997) Economic analysis of a commercial-scale aquaponic system for the production of tilapia and lettuce. In: Tilapia aquaculture: proceedings of the fourth international symposium on Tilapia in aquaculture, Orlando, Florida, pp 603–612

BORISplus (2018) Amtliche Informationen zum Immobilienmarkt in NRW. 16 Mar 2018. https://www.boris.nrw.de/borisplus/?lang=en

Bosma R (2016) The economic feasibility of aquaponics: a post-hoc cost-benefit analysis of investing in a fish vegetable farm near Dumaguete, Philippines. 11th Asian Fisheries and Aquaculture Forum, Bangkok, 2016-8-6/2016-8-6, 10 Dec 2018. http://library.wur.nl/WebQuery/wurpubs/512805

Bosma RH, Lacambra L, Landstra Y, Perini C, Poulie J, Schwaner MJ, Yin Y (2017) The financial feasibility of producing fish and vegetables through aquaponics. Aquacult Eng 78:146–154. https://doi.org/10.1016/j.aquaeng.2017.07.002

Bunyaviroch C (2013) Aquaponic systems in Puerto Rico: assessing their economic viability. PhD dissertation, Worcester Polytechnic Institute

Chryssohoidis GM, Krystallis A (2005) Organic consumers' personal values research: testing and validating the list of values (LOV) scale and implementing a value-based segmentation task. Food Qual Pref 16:585–599

Destatis (2015) Statistisches Bundesamt. https://www.destatis.de/. Accessed 10 Nov 2015

Engle CR (2015) Economics of aquaponics. SRAC Publication – Southern Regional Aquaculture Center. No.5006. 15 Feb 2018 https://srac.tamu.edu/serveFactSheet/282

Goddek S, Delaide B, Mankasingh U, Ragnarsdottir K, Jijakli H, Thorarinsdottir R (2015) Challenges of sustainable and commercial aquaponics. Sustainability 7:4199–4224. https://doi.org/10.3390/su7044199

Goddek S, Espinal C, Delaide B, Jijakli M, Schmautz Z, Wuertz S, Keesman K (2016) Navigating towards decoupled aquaponic systems: a system dynamics design approach. Water 8:303. https://doi.org/10.3390/w8070303

Graber A, Junge R (2009) Aquaponic Systems: Nutrient recycling from fish wastewater by vegetable production. Desalination 246:147–156. https://doi.org/10.1016/j.desal.2008.03.048

Hjelmar U (2011) Consumers' purchase of organic food products. A matter of convenience and reflexive practices. Appetite 56:336–334

IndustryARC (2012) Aquaponics market by equipment (grow lights, pumps, valves, greenhouse, water heaters, aeration systems); Components (rearing tanks, settling basins, sump, biofilter, hydroponics); End user (Commercial, home, community) Forecast (2013–2018), 24 Mar 2018. https://www.slideshare.net/rajesh193k/aquaponics-market-size-and-forecast-2018

IndustryARC (2017) Aquaponics market by equipment (Grow lights, pumps, valves, greenhouse, water heaters, aeration systems); Components (Rearing tanks, settling basins, sump, biofilter, hydroponics); End user (Commercial, home, community) – Forecast (2015–2021), 1 Feb 2018. http://industryarc.com/Report/22/global-commercial-aquaponics-market.html

Karimanzira D, Keesman KJ, Kloas W, Baganz D, Rauschenbach T (2016) Dynamic modeling of the INAPRO aquaponic system. Aquacult Eng 75:29–45. https://doi.org/10.1016/j.aquaeng.2016.10.004

Knaus U, Palm HW (2017) Effects of the fish species choice on vegetables in aquaponics under spring-summer conditions in northern Germany (Mecklenburg Western Pomerania). Aquaculture 473:62–73. https://doi.org/10.1016/j.aquaculture.2017.01.020

Korn A, Feucht Y, Zander K, Janssen M, Hamm U (2014): Entwicklung einer Kommunikationsstrategie für nachhaltige Aquakulturprodukte. 29 Oct 2017. http://orgprints.org/28279/1/28279-11NA040-066-uni-kassel-ti-2014-fischlabelling.pdf

Körner O, Holst N (2017) An open-source greenhouse modelling platform. Acta Horticult 241–248. https://doi.org/10.17660/ActaHortic.2017.1154.32

Körner O, Gutzmann E, Kledal PR (2017) A dynamic model simulating the symbiotic effects in aquaponic systems. Acta Horticult:309–316. https://doi.org/10.17660/ActaHortic.2017.1170.37

KTBL (2009) Gartenbau: Produktionsverfahren planen und kalkulieren. Kuratorium für Technik und Bauwesen in der Landwirtschaft. KTBL

Laidlaw J (2013) From Innovation to sustainability: Surviving five years as an urban aquaponics social enterprise. 13 Feb 2018. http://citiesprogramme.org/wp-content/uploads/2016/08/laidlaw_from_innovation_to-sustainability-web.pdf

Levin IP, Schneider SL, Gaeth GJ (1998) All frames are not created equal: a typology and critical analysis of framing effects. Organ Behav Hum Decis Process 76(2):149–188

Lewis WM, Yopp JH, Schramm HL, Brandenburg AM (1978) Use of hydroponics to maintain quality of recirculated water in a fish culture system. Trans Am Fish Soc 107:92–99. https://doi.org/10.1577/1548-8659(1978)107

Leyer M, Hüttel S (2017) Performance analysis with DEA, process mining and business process simulation on a livestock process. In: Proceedings of the international multiconference of engineers and computer scientists, 13 Dec 2018. https://pdfs.semanticscholar.org/2d2d/ef068539537127feed0080655cdba4266ce6.pdf

Lohrberg F, Lička L, Scazzosi L, Timpe A (eds) (2016) Urban agriculture Europe. Jovis, Berlin

Love DC, Fry JP et al (2014) An international survey of aquaponics practitioners. PLoS One 9(7). https://doi.org/10.1371/journal.pone.0102662

Love DC, Fry JP, Li X, Hill ES, Genello L, Semmens K, Thompson RE (2015) Commercial aquaponics production and profitability: findings from an international survey. Aquaculture 435:67–74. https://doi.org/10.1016/j.aquaculture.2014.09.023

Mackowiak C, Owens L, Hinkle C, Prince R (1989) Continuous hydroponic wheat production using a recirculating system. NASA Technical Memorandum TM 102784. 3 Mar 2018. https://ntrs.nasa.gov/archive/nasa/casi.ntrs.nasa.gov/19900009537.pdf

Medina M, Jayachandran K, Bhat MG, Deoraj A (2015) Assessing plant growth, water quality and economic effects from application of a plant-based aquafeed in a recirculating aquaponic system. Aquacult Intl 24(1):415–427. https://doi.org/10.1007/s10499-015-9934-3

Mergenthaler M, Lorleberg W (2016) Voruntersuchung zur Zahlungsbereitschaft für frischen und geräucherten europäischen Wels aus einer Aquaponik-Pilotanlage. In: Notizen aus der Forschung, Fachbereich Agrarwirtschaft, Soest,14/2016

Miličić V, Thorarinsdottir R, Santos M, Hančič M (2017) Commercial aquaponics approaching the European market: to consumers' perceptions of aquaponics products in Europe. Water 9:80. https://doi.org/10.3390/w9020080

Morgenstern R, Lorleberg W, Biernatzki R, Boelhauve M, Braun, J., Haberlah-Korr, V., 2017. Pilotstudie "Nachhaltige Aquaponik-Erzeugung für Nordrhein-Westfalen." Fachhochschule Südwestfalen, Soest. ISBN:978-3-940956-66-8

Naegel LCA (1977) Combined production of fish and plants in recirculating water. Aquaculture 10:17–24. https://doi.org/10.1016/0044-8486(77)90029-1

Orsini F, Dubbeling M, de Zeeuw H, Gianquinto G (2017) Rooftop urban agriculture. Springer

Pakseresht A, McFadden BR, Lagerkvist CJ (2017) Consumer acceptance of food biotechnology based on policy context and upstream acceptance: evidence from an artefactual field experiment. Eur Rev Agricult Econom 44:757–780. https://doi.org/10.1093/erae/jbx016

Palm HW, Knaus U, Appelbaum S, Goddek S, Strauch SM, Vermeulen T, Haïssam Jijakli M, Kotzen B (2018) Towards commercial aquaponics: a review of systems, designs, scales and nomenclature. Aquacult Intl. https://doi.org/10.1007/s10499-018-0249-z

Peattie K (1992) Green marketing: The M and E handbook series. Pitman Longman Group, London

Petrea SM, Coadă MT, Cristea V, Dediu L, Cristea D, Rahoveanu AT, Zugravu AG, Rahoveanu MMT, Mocuta DN (2016) A comparative cost – effectiveness analysis in different tested aquaponic systems. Agricult Agricult Sci Proc 10:555–565. https://doi.org/10.1016/j.aaspro.2016.09.034

Pollard G, Ward J, Koth B (2017) Aquaponics in urban agriculture: social acceptance and urban food planning. Horticulturae 3:39. https://doi.org/10.3390/horticulturae3020039

Quagrainie KK, Flores RMV, Kim H-J, McClain V (2018) Economic analysis of aquaponics and hydroponics production in the U.S. Midwest. J Appl Aquacult 30:1–14

Savidov N, Brooks A (2004) Evaluation and development of aquaponics production and product market capabilities in Alberta. Crop Diversification Centre South, Alberta Agriculture, Food and Rural Development

Schröter I, Mergenthaler M (2018a) Eye-tracking als neuroökonomische Methode zur Erfassung von Wahrnehmungsprozessen von Informationen über Aquaponik. Notizen aus der Forschung Nr. 2/2018. Fachbereich Agrarwirtschaft, Soest

Schröter I, Mergenthaler M (2018b) Technik oder Natur: Wahrnehmung und Bewertung von Aquaponik-Varianten. Notizen aus der Forschung Nr. 1/2018. Fachbereich Agrarwirtschaft, Soest

Schröter I, Hüppe J-H, Lorleberg W, Mergenthaler M (2017a) Einflussfaktoren auf die Zahlungsbereitschaft für Fisch nach dem Besuch einer Aquaponikanlage. In: Notizen aus der Forschung, Fachbereich Agrarwirtschaft, Soest.71

Schröter I, Hüppe J-H, Lorleberg W, Mergenthaler M (2017b) Kenntnisse über Aquaponik und Zahlungsbereitschaft für Fisch nach dem Besuch einer Polykulturanlage mit Aquaponik. In: Notizen aus der Forschung, Fachbereich Agrarwirtschaft, Soest. 70

Schröter I, Lorleberg W, Mergenthaler M (2017c) Erwartungen potentieller Konsumenten an Fischprodukte aus Aquaponik in einer Sondierungsstichprobe. In: Notizen aus der Forschung, Fachbereich Agrarwirtschaft, Soest,69

Schuetz K, Voigt L, Mergenthaler M (2018) Regionale Lebensmittel zwischen Anspruch und Wirklichkeit: Verbrauchererwartungen, Begriffs- und Qualitätsverständnisse regionaler Vermarktungsinitiativen und verbraucherpolitische Implikationen. KVF working paper

Smeets PJ (2010) Expedition Agroparks; Research by Design Into Sustainable Development and Agriculture in the Network Society. Wageningen Academic Publications

Smil V (2008) Energy in nature and society: general energetics of complex systems. MIT Press, Cambridge, MA

Specht K, Sanyé-Mengual E (2017) Risks in urban rooftop agriculture: Assessing stakeholders' perceptions to ensure efficient policymaking. Environ Sci Policy 69:13–21. https://doi.org/10.1016/j.envsci.2016.12.001

Specht K, Weith T, Swoboda K, Siebert R (2016) Socially acceptable urban agriculture businesses. Agron Sustain Dev 36(1):17. https://doi.org/10.1007/s13593-016-0355-0

Tamin M, Harun A, Estim A, Saufie S, Obong S (2015) Consumer Acceptance towards Aquaponic Products. IOSR J Bus Manage (IOSR-JBM) 17:49–64. https://doi.org/10.9790/487X-17824964

Thorarinsdottir RI (2015) Aquaponics guidelines, University of Iceland: Iceland. Available online: http://skemman.is/en/stream/get/1946/23343/52997/1/Guidelines_Aquaponics_20151112.pdf

Tokunaga K, Tamaru C, Ako H, Leung P (2015) Economics of small-scale commercial aquaponics in Hawaii. J World Aquacult Soc 46:20–32

Villarroel (2017) Map. Aquaponics Map (COST FA1305). https://www.google.com/maps/d/u/0/viewer?ll=35.35294037658608%2C0.45745135072172616&z=4&mid=1bjUUbCtUfE_BCgaAf7AbmxyCpT0. Accessed 8 Sep 2018

Villarroel M, Junge R, Komives T, König B, Plaza I, Bittsánszky A, Joly A (2016) Survey of Aquaponics in Europe. Water 8:468. https://doi.org/10.3390/w8100468

Vogt L, Schütz K, Mergenthaler M, (2016) Regionalisierung von Lieferketten in der Ernährungswirtschaft–Herausforderungen und Ansatzpunkte

Zugravu AG, Rahoveanu MMT, Rahoveanu AT, Khalel MS, Ibrahim MAR (2016) The perception of aquaponics products in Romania. Risk Contempo Econom 3:525–530

6

Aquaponics for the Anthropocene: Towards a 'Sustainability First' Agenda

James Gott, Rolf Morgenstern, and Maja Turnšek

Abstract 'The Anthropocene' has emerged as a unique moment in earth history where humanity recognises its devastating capacity to destabilise the planetary processes upon which it depends. Modern agriculture plays a central role in this problematic. Food production innovations are needed that exceed traditional paradigms of the Green Revolution whilst at the same time are able to acknowledge the complexity arising from the sustainability and food security issues that mark our times. Aquaponics is one technological innovation that promises to contribute much towards these imperatives. But this emergent field is in an early stage that is characterised by limited resources, market uncertainty, institutional resistance and high risks of failure—a developmental environment where hype prevails over demonstrated outcomes. Given this situation, the aquaponics research community potentially holds an important place in the development path of this technology. But the field needs to craft a coherent and viable vision for this technology that can move beyond misplaced techno-optimist accounts. Turning to sustainability science and STS research, we discuss the urgent need to develop what we call a 'critical sustainability knowledge' for aquaponics, giving pointers for possible ways forward, which include (1) expanding aquaponic research into an interdisciplinary research domain, (2) opening research up to participatory approaches in real-world contexts and (3) pursuing a solution-oriented approach for sustainability and food security outcomes.

Keywords Anthropocene · Green Revolution · Techno-optimism · STS research (Science, Technology and Society research)

J. Gott (✉)
Geography and Environment, University of Southampton, Southampton, UK
e-mail: j.gott@soton.ac.uk

R. Morgenstern
Department of Agriculture, University of Applied Sciences of South Westphalia, Soest, Germany
e-mail: morgenstern.rolf@fh-swf.de

M. Turnšek
Faculty of Tourism, University of Maribor, Brežice, Slovenia
e-mail: maja.turnsek@um.si

Introduction

Key drivers stated for aquaponic research are the global environmental, social and economic challenges identified by supranational authorities like the Food and Agriculture Organization (FAO) of the United Nations (UN) (DESA 2015) whose calls for sustainable and stable food production advance the 'need for new and improved solutions for food production and consumption' (1) (Junge et al. 2017; Konig et al. 2016). There is growing recognition that current agricultural modes of production cause wasteful overconsumption of environmental resources, rely on increasingly scarce and expensive fossil fuel, exacerbate environmental contamination and ultimately contribute to climate change (Pearson 2007). In our time of 'peak-everything' (Cohen 2012), 'business as usual' for our food system appears at odds with a sustainable and just future of food provision (Fischer et al. 2007). A food system revolution is urgently needed (Kiers et al. 2008; Foley et al. 2011), and as the opening chapters of this book attest, aquaponics technology shows much promise. The enclosed systems of aquaponics offer an especially alluring convergence of potential resolutions that could contribute towards a more sustain-able future (Kőmíves and Ranka 2015). But, we ask, what kind of sustainable future might aquaponics research and aquaponics technology contribute towards? In this chapter, we take a step back to consider the ambitions of our research and the functions of our technology.

In this chapter we situate current aquaponic research within the larger-scale shifts of outlook occurring across the sciences and beyond due to the problematic that has become known as 'the Anthropocene' (Crutzen and Stoermer 2000b). Expanding well beyond the confines of its original geological formulation (Lorimer 2017), the Anthropocene concept has become no less than 'the master narrative of our times' (Hamilton et al. 2015). It represents an urgent realisation that demands deep questions be asked about the way society organises and relates to the world, including the *modus operandi* of our research (Castree 2015). However, until now, the concept has been largely sidelined in aquaponic literature. This chapter introduces the Anthropocene as an obligatory frame of reference that must be acknowledged for any concerted effort towards future food security and sustainability.

We discuss how the Anthropocene unsettles some key tenets that have underpinned the traditional agriscience of the Green Revolution (Stengers 2018) and how this brings challenges and opportunities for aquaponic research. Aquaponics is an innovation that promises to contribute much towards the imperatives of sustainability and food security. But this emergent field is in an early stage that is characterised by limited resources, market uncertainty, institutional resistance with high risks of failure and few success stories—an innovation environment where hype prevails over demonstrated outcomes (König et al. 2018). We suggest this situation is characterised by a misplaced techno-optimism that is unconducive to the deeper shifts towards sustainability that are needed of our food system.

Given this, we feel the aquaponics research community has an important role to play in the future development of this technology. We suggest a refocusing of aquaponics research around the key demands of our food system—sustainability and food security. Such a task entails we more thoroughly consider the nature of sustainability, and so we draw on the insights from the fields of sustainability science and STS. Addressing sustainability in the Anthropocene obligates the need to attend more holistically the interacting biophysical, social, economic, legal and ethical dimensions that encroach on aquaponic systems (Geels 2011). This is no small task that places great demands on the way we produce and use knowledge. For this reason we discuss the need to develop what we call a 'critical sustainability knowledge' for aquaponics, giving pointers for possible ways forward, which include (1) expanding aquaponic research into an interdisciplinary research domain, (2) opening research up to participatory approaches in real-world contexts and (3) pursuing a solution-oriented approach for sustainability and food security outcomes.

The Anthropocene and Agriscience

'Today, humankind has begun to match and even exceed some of the great forces of nature [...] [T]he Earth System is now in a no analogue situation, best referred to as a new era in the geological history, the Anthropocene' (Oldfield et al. 2004: 81).

The scientific proposal that the Earth has entered a new epoch—'the Anthropocene'—as a result of human activities was put forward at the turn of the new millennium by the chemist and Nobel Laureate Paul Crutzen and biologist Eugene Stoermer (Crutzen and Stoermer 2000a). Increasing quantitative evidence suggests that anthropogenic material flows stemming from fossil fuel combustion, agricultural production and mineral extraction now rival in scale those natural flows supposedly occurring outside of human activity (Steffen et al. 2015a). This is a moment marked by unprecedented and unpredictable climatic, environmental and ecological events (Williams and Jackson 2007). The benign era of the Holocene has passed, so the proposal claims; we have now entered a much more unpredictable and dangerous time where humanity recognises its devastating capacity to destabilise planetary processes upon which it depends (Rockström et al. 2009, Steffen et al. 2015b; See chapter). The Anthropocene is therefore a moment of realisation, where the extent of human activities must be reconciled within the boundaries of biophysical processes that define the safe operating space of a stable and resilient Earth system (Steffen et al. 2015b).

A profound intertwining of the fates of nature and humankind has emerged (Zalasiewicz et al. 2010). The growing awareness of environmental and human calamity—and our belated, tangled role within it—puts to test our faith in the key modernist assumption, namely, the dualisms separating humans from nature (Hamilton et al. 2015). This is a shocking and unprecedented moment because modernist epistemologies have proven exceedingly powerful, contributing significantly towards the organisation of society to the present day (Latour 1993). Conceptions of unique and stable human agency, the presumption of progressive norms such as liberty or universal dignity, and the existence of an objective world separate from human doings are all put to test (Latour 2015; Hamilton et al. 2015).

This insight, without doubt, applies to the food system of which we all inherit. The Green Revolution[1] was underpinned with modern aspirations, being founded on ideas such as linear notions of progress, the power of human reason and faith in the inevitable technological resolution of human problems (Cota 2011). These conceptions, which have traditionally secured the role of science in society, begin to appear increasingly unreliable with the advent of the Anthropocene (Savransky 2013; Stengers 2015). The inconvenient truth is that the technoscientific interventions, which have been implemented as modern agrarian solutions onto our world over the last century, have carried with them serious and unexpected outcomes. What's more, these escalating biophysical disruptions (e.g. greenhouse gas emissions and nitrogen and phosphorous cycle perturbations) that have only recently become perceived must be added to a much broader series of environmental, biological and social repercussions brought about by particular aspects of our modernised food system.

The Anthropocene problematic leaves little doubt that our contemporary food system faces enormous challenges (Kiers et al. 2008; Baulcombe et al. 2009; Pelletier and Tyedmers 2010). Prominent studies point to agriculture as the single largest contributor to the rising environmental risks posed in the Anthropocene (Struik and Kuyper 2014; Foley et al. 2011). Agriculture is the single largest user of freshwater in the world (Postel 2003); the world's largest contributor to altering the global nitrogen and phosphorus cycles and a significant source (19–29%) of greenhouse gas emissions (Vermeulen et al. 2012; Noordwijk 2014). Put simply, 'agriculture is a primary driver of global change' (Rockström et al. 2017:6). And yet, it is from within the new epoch of the Anthropocene that the challenge of feeding humanity must be resolved. The number of hungry people in the world persists at approximately 900 million (FAO, Ifad and WFP. 2013). Even then, in order to feed the world by 2050, best estimates suggest that production must roughly double to keep pace with projected demands from population growth, dietary changes (particularly meat consumption) and increasing bioenergy use (Kiers et al. 2008; Baulcombe et al. 2009; Pelletier and Tyedmers 2010; Kearney 2010). Complicating matters even further is the need not simply to produce more, but also to manage the entire food system more efficiently. In a world where 2 billion suffer from micronutrient deficiencies, whilst 1.4 billion adults are over-nourished, the need for better distribution, access and nutrition is glaring, as is the drastic need to reduce the deplorable levels of waste (conservative estimates suggest 30%) in the farm-to-fork supply chain (Parfitt et al. 2010; Lundqvist et al. 2008; Stuart 2009).

---

[1]The Green Revolution refers to a set of research and technology transfer initiatives occurring from the 1930s and the late 1960s that increased agricultural production worldwide, particularly in the developing world. As Farmer (1986) describes, these initiatives resulted in the adoption of new technologies, including: 'New, high-yielding varieties of cereals... in association with chemical fertilizers and agro-chemicals, and with controlled water-supply... and new methods of cultivation, including mechanization. All of these together were seen as a "package of practices" to supersede "traditional" technology and to be adopted as a whole'.

The Anthropocene problematic presents serious questions about modern industrial agriculture, which in many guises is now deemed inefficient, destructive and inadequate for our new global situation. But the fallout of this situation is more considerable still, for the Anthropocene strikes a challenge at the very agricultural paradigm currently dominating food provision (Rockström et al. 2017). For this reason the challenge extends well beyond 'the farm' and incorporates a much wider set of structures, practices and beliefs that continue to enact and propel the modern agricultural paradigm into our newly demanding epoch. With this comes the urgent need to reconsider the methods and practices, ambitions and goals that define our current agriscience research. Are they fit for the challenges of our new epoch, or do they merely reproduce inadequate visions of modernist food provision?

Getting Beyond the Green Revolution

The Anthropocene marks a step change in the relation between humans and our planet. It demands a rethink of the current modes of production that currently propel us on unsustainable trajectories. Until now, such reflexive commitments have not been required of agriscience research and development. It is worth remembering that the Green Revolution, in both its ambitions and methods, was for some time uncontroversial; agriculture was to be intensified and productivity per unit of land or labour increased (Struik 2006). Without doubt, this project, whose technological innovations were vigorously promoted by governments, companies and foundations around the world (Evenson and Gollin 2003), was phenomenally successful across vast scales. More calories produced with less average labour time in the commodity system was the equation that allowed the cheapest food in world history to be produced (Moore 2015). In order to simplify, standardise and mechanise agriculture towards increases in productivity per worker, plant and animal, a series of biophysical barriers had to be overridden. The Green Revolution achieved this largely through non-renewable inputs.

In the Anthropocene, this agricultural paradigm that marked the Green Revolution runs up against (geological) history. Growing awareness is that this 'artificialised' agricultural model, which substitutes each time more ecological processes with finite chemical inputs, irrigation and fossil fuel (Caron et al. 2014), literally undermines the foundations of future food provision. The biophysical contradictions of late-capitalist industrial agriculture have become increasingly conspicuous (Weis 2010). Moreover, the dramatic environmental, economic and social consequences of contemporary models of high-intensity artificialised agriculture have become an escalating concern for a globalised food system manifesting accelerating contradictions (Kearney 2010; Parfitt et al. 2010).

During the post-war period (mid-40s–70s), secure economic growth was founded on the accelerated extraction of fossil fuel, and as Cota (Cota 2011) notes, agriscience development during this time progressed more in tune with the geochemical sciences than the life sciences. Agricultural production designed around the cheapest maximum yields had been simplified and unified into monocrops, made to depend on mechanisation and agrochemical products. Although highly effective when first implemented, the efficiency of these commercial inputs has witnessed diminishing returns (Moore 2015). Following the oil crises of the 70s, the productivist ideals of the Green Revolution fell more upon the life sciences, particularly in the guise of agri-biotech, which has grown into a multibillion-dollar industry.

Feeding the globe's exploding population has been the key concern in a decade-long productivist narrative that has served to secure the prominent position of agricultural biotech in our current food system (Hunter et al. 2017). The great shock is that this highly advanced sector has done little to improve intrinsic yields. World agricultural productivity growth slowed from 3% a year in the 1960s to 1.1% in the 1990s (Dobbs et al. 2011). Recently, the yields of key crops have in some places approached plateaux in production (Grassini et al. 2013). Mainstream agroscientists have voiced concern that the maximum yield potential of current varieties is fast approaching (Gurian-Sherman 2009). On top of this, climate change is estimated to have already reduced global yields of maize and wheat by 3.8% and 5.5%, respectively (Lobell et al. 2011), and some warn of sharp declines in crop productivity when temperatures exceed critical physiological thresholds (Battisti and Naylor 2009).

The waning efficiency gains of artificial inputs added to the biological limits of traditional varieties is a situation that, for some, further underscores the need to accelerate the development of genetically engineered varieties (Prado et al. 2014). Even then, the greatest proponents of GM—the biotech firms themselves—are aware that GM interventions rarely work to increase yield, but rather to *maintain* it through pesticide and herbicide resistance (Gurian-Sherman 2009). As such, agricultural production has become locked into a cycle that requires the constant replacement of new crop varieties and product packages to overcome the growing negative environmental and biological impingements upon yield [2]. Melinda Cooper's (2008: 19) influential analysis of agro-biotechnology has traced how neoliberal modes of production become relocated ever more within the genetic, molecular and cellular levels. As such, the commercialisation of agrarian systems increasingly extends towards the capture of germplasm and DNA, towards 'life itself' (Rose 2009). Cooper's (2008) diagnosis is that we are living in an era of capitalist delirium characterised by its attempt to overcome biophysical limits of our earth through the speculative biotechnological reinvention of the future. In this respect, some have argued that rather than overcoming weaknesses of the conventional paradigm, the narrow focus of GM interventions seems only to intensify its central characteristics (Altieri 2007).

Amidst the deceleration of yield increases, the estimated targets of 60–100% increases in production needed by 2050 (Tilman et al. 2011; Alexandratos and Bruinsma 2012) appear increasingly daunting. As compelling and clear as these targets may be, concerns have been raised that productivist narratives have eclipsed other pressing concerns, namely, the environmental sustainability of production (Hunter et al. 2017) and food security (Lawrence et al. 2013). The current agricultural paradigm has held production first and sustainability as a secondary task of mitigation (Struik et al. 2014).

Thirty years of frustrated sustainability talk within the productivist paradigm are testament to the severe difficulties for researchers and policymakers alike to bridge the gap between sustainability theory and practice (Krueger and Gibbs 2007). 'Sustainability' as a concept had initially had revolutionary potential. Key texts such as the Club of Rome's *The Limits of Growth* (Meadows et al. 1972), for instance, contained an imminent critique of global development narratives. But researchers have pointed out the way that 'sustainability' throughout the 80s and 90s became assimilated into neoliberal growth discourse (Keil 2007). We now have a situation where, on the one hand, global sustainability is almost unanimously understood as a prerequisite to attain human development across all scales—from local, to city, nation and the world (Folke et al. 2005)—whilst on the other, despite substantial efforts in many levels of society towards the creation of a sustainable future, key global-scale indicators show that humanity is actually moving away from sustainability rather than towards it (Fischer et al. 2007). This is in spite of the increasing regularity of high-profile reports that evermore underscore the grave risks of existing trends to the long-term viability of ecological, social and economic systems (Steffen et al. 2006; Stocker 2014; Assessment 2003; Stern 2008). This situation—the widening gap between our current trajectory and all meaningful sustainability targets—has been discussed as the so-called 'paradox of sustainability' (Krueger and Gibbs 2007). Prevailing discourse on food security and sustainability continues to galvanise growth-oriented developmental imperatives (Hunter et al. 2017).

Agriscience research and development proliferated in accordance with the dominant politico-economic structures that defined planetary development over the last 30 years (Marzec 2014). Although the negative effects of the so-called 'Chicago School' of development have by now been well documented (Harvey 2007), biotechnological innovation remains rooted within neoliberal discourse (Cooper 2008). These narratives consistently present global markets, biotech innovation and multinational corporate initiatives as the structural preconditions for food security and sustainability. The empirical credibility of such claims has long been challenged (Sen 2001), but seem especially relevant amidst the accumulating history of chronic distributional failures and food crises that mark our times. It is worth repeating Nally's (2011; 49) point: 'The spectre of hunger in a world of plenty seems set to continue into the 21st century... this is not the failure of the modern food regime, but the logical expression of its central paradoxes'. The situation is one where malnutrition is seen no longer as a failure of an otherwise efficiently functioning system, but rather as an endemic feature within the systemic production of scarcity (Nally 2011). In the face of such persisting inconsistencies, commentators note that neoliberal appeals to human prosperity, food security and green growth appear out of touch and often ideologically driven (Krueger and Gibbs 2007).

The Anthropocene is a time where ecological, economic and social disaster walk hand in hand as modern economies and institutions geared towards unlimited growth crash against the finite biophysical systems of the earth (Altvater et al. 2016; Moore

2015). Cohen (2013) describes the Anthropocene as an 'eco-eco' disaster, paying heed to the rotten relationship in which eco*nomic* debt becomes compounded against the eco*logical* debt of species extinction. Now more than ever, faith in the modernising powers of neoliberal food interventions proclaiming just and sustainable futures wears thin (Stengers 2018), yet the resemblance noted by some commentators (Gibson-Graham 2014), between our food system and the unhinged financial systems of our neoliberal economies charts an alarming trend. It's worth noting this resemblance runs deeper than the mere production of debt (one being calorific and genetic, the other economic). The truth is our food system hinges on a cash nexus that links trade tariffs, agricultural subsidies, enforcement of intellectual property rights and the privatisation of public provisioning systems. Viewed from above, these procedures constitute a pseudo-corporate management of the food system, which according to Nally (2011: 37) should be seen as a properly *biopolitical* process designed for managing life, "including the lives of the hungry poor who are 'let die' as commercial interests supplant human needs". Petrochemicals and micronutrients, it seems, are not the only things being consumed in the Anthropocene; futures are (Collings 2014; Cardinale et al. 2012).

What once might have been considered necessary side effects of the modernising imperative of the Green Revolution, the so-called 'externalities' of our current food system, are increasingly exposed as a kind of 'deceptive efficiency' bent towards rapid production and profit and very little else (Weis 2010). The disturbing realisation is that the food system we inherit from the Green Revolution creates value only when a great number of costs (physical, biological, human, moral) are allowed to be overlooked (Tegtmeier and Duffy 2004). A growing number of voices remind us that costs of production go beyond the environment into matters such as the exclusion of deprived farmers, the promotion of destructive diets (Pelletier and Tyedmers 2010) and more generally the evacuation of social justice and political stability from matters of food provision (Power 1999). The relation between agrarian technological intervention, food security and sustainability emerges as far wider and complex issue than could be acknowledged by narratives of the Green Revolution.

Situating the contemporary food system within dominant recent historical processes, the above discussion has paid particular attention to destructive links between modern agriculture and the economic logics of late capitalism. It is important, however, to remember that numerous commentators have cautioned against oversimplified or deterministic accounts regarding the relationship between capitalist relations of production and the Anthropocene problematic (Stengers 2015; Haraway 2015; Altvater et al. 2016). Such a discussion is made possible by close to four decades of critical investigations by feminists, science and technology scholars, historians, geographers, anthropologists and activists, which have endeavoured in tracing the links between hegemonic forms of science and the social/environmental destruction caused by industrial capitalism (Kloppenburg 1991). This 'deconstructive' research ethic developed important understandings of the way modern agriscience progressed down trajectories that involve the neglect of particular physical, biological, political and social contexts and histories (Kloppenburg 1991). In many instances, the modernising narratives of 'development' like those put to work in the Green Revolution became seen—by anthropologists, historians and indigenous communities alike—as a kind of modified successor to pre-war colonial discourse (Scott 2008; Martinez-Torres and Rosset 2010). In anthropological terms, what these studies taught us was that although modern agriculture was rooted in developmental narratives of universal prosperity, in reality, 'progress' was achieved through the displacement or indeed destruction of a great diversity of agricultural perspectives, practices, ecologies and landscapes. It is for this reason Cota (2011: 6) reminds us of the importance of the critical work that explicitly positioned the biopolitical paradigm of industrial agriculture 'not first and foremost as an economic kind of imperialism, but more profoundly as an epistemic and culturally specific kind of imperialism'.

This is a key point. The Green Revolution was not merely a technical, nor economic intervention, but involved the spread of a more profound reconfiguration of the epistemological registers of food provision itself. It was a process that deeply influenced the way agricultural knowledge was produced, propagated and implemented. As Cota (2011: 6) explains: 'the use of physicalist and probabilistic discourse, a purely instrumental conception of nature and work, the implementation of statistical calculations disconnected from local conditions, [as well as] the reliance on models without recognizing historic specificities' were all ways of enacting the biopolitical agenda of the Green Revolution. This list of commitments describes the fundamentals at the sharp end of the Green Revolution, but as we have seen, such commitments alone have proven insufficient for the task of creating a just and sustainable food system. It becomes apparent that any research agenda fit for the Anthropocene must learn to move beyond the modern food paradigm by forging a different research ethic with different commitments.

Paradigm Shift for a New Food System

To claim that Agriculture is 'at a crossroads' (Kiers et al. 2008) does not quite do justice to the magnitude of the situation. The gaping 'sustainability gap' (Fischer et al. 2007) amidst unanimous calls for sustainability are increasingly being met with common response amongst researchers: pleas for revolutionary measures and paradigm shifts. Foley et al. (2011: 5) put it quite directly: 'The challenges facing agriculture today are unlike anything we have experienced before, and they require revolutionary approaches to solving food production and sustainability problems. In short, new agricultural systems must deliver more human value, to those who need it most, with the least environmental harm'. Somehow, world agriculture's current role as the single largest driver of global environmental change must shift into a 'critical agent of a world transition' towards global sustainability within the biophysical safe operating space of the Earth (Rockström et al. 2017).

The Anthropocene lays steep demands: Agriculture must be intensified; it must meet the needs of a growing population, but at the same time it is mandatory that the pressures exerted by our food production systems stay within the carrying capacity of Planet Earth. It is increasingly understood that future food security depends on the development of technologies that increase the efficiency of resource use whilst simultaneously preventing the externalisation of costs (Garnett et al. 2013). The search for alternatives to our current agricultural paradigm has brought to the fore ideas such as agroecology (Reynolds et al. 2014) and 'sustainable intensification', with the acknowledgement that real progress must be made towards 'ecological intensification', that is, increasing agricultural output by capitalising on the ecological processes in agroecosystems (Struik and Kuyper 2014).

There has been well-documented debate on what constitutes 'sustainable intensification' (SI) of agriculture as well as the role it might play in addressing global food security (Struik and Kuyper 2014; Kuyper and Struik 2014; Godfray and Garnett 2014). Critics have cautioned against the top-down, global analyses that are often framed in narrow, production-oriented perspectives, calling for a stronger engagement with the wider literature on sustainability, food security and food sovereignty (Loos et al. 2014). Such readings revisit the need for developing regionally grounded, bottom-up approaches, with a growing consensus claiming that an SI agenda fit for the Anthropocene does not entail 'business-as-usual' food production with marginal improvements in sustainability but rather a radical rethinking of food systems not only to reduce environmental impacts but also to enhance animal welfare, human nutrition and support rural/urban economies with sustainable development (Godfray and Garnett 2014).

While traditional 'sustainable intensification' (SI) has been criticised by some as too narrowly focused on production, or even as a contradiction in terms altogether (Petersen and Snapp 2015), others make it clear that the approach must be broadly conceived, with the acknowledgement that there is no single universal pathway to sustainable intensification (Garnett and Godfray 2012). Important here is the growing appreciation of 'multifunctionality' in agriculture (Potter 2004). If, during the twentieth century, 'Malthusian' demographics discourse had secured the narrow goal of agricultural development on increasing production, the growing rediscovery of the multiple dimensions of farming currently taking place is altering the perception of the relationship between agriculture and society.

'Multifunctionality' as an idea was initially contested in the context of the controversial GATT and WTO agricultural and trade policy negotiations (Caron et al. 2008), but has since gained wide acceptance, leading to a more integrative view of our food system (Potter 2004). In this view, progress in seeing agriculture as an important type of 'land use' competing with other land functions (Bringezu et al. 2014) interrelates with a number of other perspectives. These have been conceptualised through several important categories: (1) as a source of employment and livelihood for a rural and future urban population (McMichael 1994); (2) as a key part of cultural heritage and identity (van der Ploeg and Ventura 2014); (3) as the basis of complex value chain interactions in 'food systems' (Perrot et al. 2011); (4) as a sector in regional, national and global economies (Fuglie 2010); (5) as modifier and storehouse of genetic resources (Jackson et al. 2010); (6) as a threat to environmental integrity that exerts destructive pressures on biodiversity (Brussaard et al. 2010; Smil 2011); and (7) as a source of greenhouse gas emissions (Noordwijk

2014). This list is by no means comprehensive, but what is important is that each of these interacting dimensions is understood to impact sustainability and food security in one way or another and must be apprehended by serious attempts towards SI.

Sustainability outcomes are increasingly seen as a complex interplay between local and global concerns (Reynolds et al. 2014). Biophysical, ecological and human needs intermix within the complexities and idiosyncrasies of 'place' (Withers 2009). The 'one size fits all' solutions, characteristics of the Green Revolution, fail to acknowledge these unique sustainability potentials and demands. The result is that changes in food production and consumption must be perceived through a multiplicity of scales *and* styles. To this end, Reynolds et al. (2014) suggest an approach to sustainability that takes advantage of the insights of agroecological principles. They forward a 'custom-fit' food production focus 'explicitly tailored to the environmental and cultural individuality of place and respectful of local resource and waste assimilative limits, thus promoting biological and cultural diversity as well as steady-state economics'.

If the issues at stake are inherently *multidimensional*, others have also underlined that they are *contested*. Trade-offs between the plethora of biophysical and human concerns are inevitable and often exceedingly complex. Sustainability thresholds are diverse, often normative, and can seldom all be realised in full simultaneously (Struik and Kuyper 2014). It has been emphasised that new directions towards sustainability and food security require simultaneous change at the level of formal and informal social rules and incentive systems (i.e. institutions) that orient human interaction and behaviour, and hence that 'institutional innovation' is held to be a key entry point in addressing challenges (Hall et al. 2001). Insomuch as the complexity of sustainable intensification derives from human framings (which entail and flow from contexts, identities, intentions, priorities and even contradictions), they are, as Kuyper and Struik (2014: 72) put it, 'beyond the command of science'. Attempting to reconcile the many dimensions of food production towards sustainable ends and within the bounds of our finite planet involves a great deal of uncertainty, irreducibility and contestation (Funtowicz and Ravetz 1995); it requires an awareness and acknowledgement that such issues are shot through with *political* implication.

Food systems and sustainability research have come a long way in expanding the narrow focus of the Green Revolution, bringing greater clarity to the steep challenges we face in the pursuit of a more environmentally and socially sustainable food system. Thanks to a broad range of work, it is now apparent that food production lies at the heart of a nexus of interconnected and multi-scalar processes, on which humanity relies upon to meet a host of multidimensional—often contradictory—needs (physical, biological, economic, cultural). As Rockström et al. (2017: 7) have stated: 'World agriculture must now meet social needs and fulfil sustainability criteria that enables food and all other agricultural ecosystem services (i.e., climate stabilization, flood control, support of mental health, nutrition, etc.) to be generated within a safe operating space of a stable and resilient Earth system'. It is precisely within these recalibrated agricultural goals that aquaponics technology must be developed.

Aquaponic Potential or Misplaced Hope?

Contemporary aquaponic research has shown keen awareness of particular concerns raised in the Anthropocene problematic. Justifications for aquaponic research have tended to foreground the challenge of food security on a globe with an increasing human population and ever strained resource base. For instance, König et al. (2016) precisely situate aquaponics within the planetary concerns of Anthropocene discourse when they state: 'Assuring food security in the twenty-first century within sustainable planetary boundaries requires a multi-faceted agro-ecological intensification of food production and the decoupling from unsustainable resource use'. Towards these important sustainability goals, it is claimed that aquaponic technology shows much promise (Goddek et al. 2015). The innovative enclosed systems of aquaponics offer an especially alluring convergence of potential resolutions that could contribute towards a more sustainable future.

Proponents of aquaponics often stress the ecological principles at the heart of this emerging technology. Aquaponic systems harness the positive potential of a more or less simple ecosystem, in order to reduce the use of finite inputs whilst simultaneously reducing waste by-products and other externalities. On these grounds, aquaponic technology can be viewed as a primary example of 'sustainable intensification' (Garnett et al. 2013) or, more precisely, as a form of 'ecological intensification' since its founding principles are based on the management of service-providing organisms towards quantifiable and direct contributions to agricultural production (Bommarco et al. 2013). From this agroecological principle flow a great number of potential sustainability benefits. Chapters of this book do an exemplary job of highlighting these, detailing the challenges faced by our food system and situating aquaponics science as the potential locus for a range of sustainability and food security interventions. There is no need to repeat these points again, but it is worth noting this perceived convergence of potential resolutions is what drives research and strengthens the 'conviction that this technology has the potential to play a significant role in food production in the future' (7) (Junge et al. 2017).

However, despite the considerable claims made by its proponents, the future of aquaponics is less than certain. Just what kind of role aquaponics might play in transitions to sustainable food provision is still largely up for debate—crucially, we must stress, *the publication of sustainability and food security outcomes of aquaponic systems remain conspicuous by their absence across Europe* (König et al. 2018). On paper, the 'charismatic' attributes of aquaponics ensure that it can easily be presented as a 'silver bullet' type of innovation that gets to the heart of our food system's deepest sustainability and food security issues (Brooks et al. 2009). Such images have been able to garner considerable attention for aquaponics far beyond the confines of academic research—consider, for instance, the significant production of online aquaponic 'hype' in comparison to similar fields, usefully pointed out by Junge et al. (2017). It is here we may take time to point out the relationship between the perceived potential of aquaponics and 'techno-optimism'.

The introduction of every new technology is accompanied by myths that spur further interest in that technology (Schoenbach 2001). Myths are circulated amongst early adopters and are picked up by the general media often long before the scientific community has time to thoroughly analyse and answer to their claims. Myths, as Schoenbach states (2001, 362), are widely believed because they 'comprise a clear-cut and convincing explanation of the world'. These powerful explanations are able to energise and align individual, community and also institutional action towards particular ends. The 'beauty' of aquaponics, if we can call it that, is that the concept can often render down the complexity of sustainability and food security issues into clear, understandable and scalable systems metaphors. The ubiquitous image of the aquaponic cycle—water flowing between fish, plants and bacteria—that elegantly resolves food system challenges is exemplary here. However, myths on technology, whether optimistic or pessimistic, share a techno-deterministic vision of the relation between technology and society (Schoenbach 2001). Within the techno-deterministic vision of technology, it is the technology that causes important changes in society: if we manage to change the technology, we thus manage to change the world. Regardless whether the change is for the better (techno-optimism) or the worse(technophobia), the technology by itself creates an effect.

Techno-determinist views have been thoroughly critiqued on sociological, philosophical (Bradley 2011), Marxist (Hornborg 2013), material-semiotic (Latour 1996) and feminist (Haraway 1997) grounds. These more nuanced approaches to technological development would claim that technology by itself does not bring change to society; it is neither inherently good nor bad but is always embedded within society's structures, and it is those structures that enable the use and effect of the technology in question. To one degree or another, technology is an emergent entity, the effects of which we cannot know in advance (de Laet and Mol 2000). This might seem like an obvious point, but techno-determinism remains a strong, if often latent, feature within our contemporary epistemological landscape. Our innovation-driven, technological societies are maintained by discursive regimes that hold on to the promise of societal renewal through technological advancement (Lave et al. 2010). Such beliefs have been shown to have an important normative role within expert communities whether they be scientists, entrepreneurs or policymakers (Franklin 1995; Soini and Birkeland 2014).

The rise of aquaponics across Europe is intertwined with specific interests of various actors. We can identify at least five societal processes that led to the development of aquaponics: (a) interest of public authorities in funding high-tech solutions for problems of sustainability; (b) venture capital financing, motivated by the successes in IT startups, looking for 'the next big thing' that will perhaps discover the new 'unicorn' (startup companies valued at over $1 billion); (c) mass media event-focused interest in snapshot reporting on positive stories of new aquaponics startups, fuelled by the public relations activities of these startups, with rare media follow-up reporting on the companies that went bust; (d) internet-supported growth of enthusiastic, do-it-yourself aquaponics communities, sharing both sustainability values and love for tinkering with new technology; (e) interests of urban developers to find economically viable solutions for vacant urban spaces and greening of urban space; and (f) research communities focused on developing technological solutions to impending sustainability and food security problems. To a greater or lesser degree, the spectre of techno-optimistic hope permeates the development of aquaponics.

Although the claims of techno-optimist positions are inspiring and able to precipitate the investment of money, time and resources from diverse actors, the potential for such standpoints to generate justice and sustainability has been questioned on scales from local (Leonard 2013) and regional issues (Hultman 2013) to global imperatives (Hamilton 2013). And it is at this point, we might consider the ambitions of our own field. A good starting point would be the 'COST action FA1305', which has been an important facilitator of Europe's aquaponic research output over recent years, with a number of publications acknowledging the positive impact of the action in enabling research (Miličić et al. 2017; Delaide et al. 2017; Villarroel et al. 2016). Like all COST actions, this EU-funded transnational networking instrument has acted as a hub for aquaponic research in Europe, galvanising and broadening the traditional networks amongst researchers by bringing together experts from science, experimental facilities and entrepreneurs. The original mission statement of COST action FA1305 reads as follows:

> *Aquaponics has a key role to play in food provision and tackling global challenges such as water scarcity, food security, urbanization, and reductions in energy use and food miles.* The EU acknowledges these challenges through its Common Agriculture Policy and policies on Water Protection, Climate Change, and Social Integration. A European approach is required in the globally emerging aquaponics research field building on the foundations of Europe's status as a global centre of excellence and technological innovation in the domains of aquaculture and hydroponic horticulture, The EU Aquaponics Hub aims to the development of aquaponics in the EU, by leading the research agenda through the creation of a networking hub of expert research and industry scientists, engineers, economists, aquaculturists and horticulturalists, and contributing to the training of young aquaponic scientists. The EU Aquaponics Hub focuses on three primary systems in three settings; (1) "cities and urban areas" – urban agriculture aquaponics, (2) "developing country systems" – devising systems and technologies for food security for local people and (3) "industrial scale aquaponics" – *providing competitive systems delivering cost effective, healthy and sustainable local food in the EU.* (http://www.cost.eu/COST_Actions/fa/FA1305, 12.10.2017, emphasis added).

As the mission statement suggests, from the outset of COST action FA1305, high levels of optimism were placed on the role of aquaponics in tackling sustainability and food security challenges. The creation of the COST EU Aquaponics Hub was to 'provide a necessary forum for 'kick-starting' aquaponics as a serious and potentially viable industry for sustainable food production in the EU and the world' (COST 2013). Indeed, from the authors' own participation within COST FA1305, our lasting experience was without doubt one of being part of a vibrant, enthused and highly skilled research community that were more or less united in their ambition to make aquaponics work towards a more sustainable future. Four years down the line since the Aquaponic Hub's mission statement was issued, however, the sustainability and food security potential of aquaponics remains just that—potential. At present it is uncertain what precise role aquaponics can play in Europe's future food system (König et al. 2018).

The commonly observed narrative that aquaponics provides a sustainable solution to the global challenges agriculture faces unveils a fundamental misconception of what it is actually capable to achieve. The plant side of aquaponics is horticulture, not agriculture, producing vegetables and leafy greens with high water content and low nutritional value compared to the staple foods agriculture on farmland produces. A quick comparison of current agricultural area, horticultural area and protected horticultural area, 184.332 km^2, 2.290 km^2 (1,3%) and 9,84 km^2 (0,0053%), in Germany, reveals the flaw in the narrative. Even if considering a much higher productivity in aquaponics through the utilisation of controlled environment systems, aquaponics is not even close to having the potential to make a real impact on agricultural practice. This becomes even more obvious when the ambition to be a 'food system of the future' ends in the quest for high-value crops (e.g. microgreens) that can be marketed as gourmet gastronomy.

It is well known that the development of sustainable technology is characterised by uncertainties, high risks and large investments with late returns (Alkemade and Suurs 2012). Aquaponics, in this regard, is no exception; only handful commercially operating systems exist across Europe (Villarroel et al. 2016). There appears considerable resistance to the development of aquaponic technology. Commercial projects have to contend with comparatively high technological and management complexity, significant marketing risks, as well as an uncertain regulatory situation that until now persists (Joly et al. 2015). Although it is difficult to pin down the rate of startup failure, the short history of commercial aquaponics across Europe might well be summed up as 'Small successes and big failures' (Haenen 2017). It is worth pointing out also that the pioneers already involved in aquaponics at the moment across Europe are unclear if their technology is bringing about any improvements in sustainability (Villarroel et al. 2016). Recent analysis from König et al. (2018) has shown how the challenges to aquaponics development derive from a host of structural concerns, as well as the technology's inherent complexity. Combined, these factors result in a high-risk environment for entrepreneurs and investors, which has produced a situation whereby startup facilities across Europe are forced to focus on production, marketing and market formation over the delivery of sustainability credentials (König et al. 2018). Aside from the claims of great potential, the sombre reality is that it remains to be seen just what impact aquaponics can have on the entrenched food production and consumption regimes operating in contemporary times. The place for aquaponic technology in the transition towards more sustainable food systems, it seems, has no guarantee.

Beyond the speculation of techno-optimism, aquaponics has emerged as a highly complex food production technology that holds potential but is faced with steep challenges. In general, there exists a lack of knowledge about how to direct research activities to develop such technologies in a way that preserves their promise of sustainability and potential solutions to pressing food system concerns (Elzen et al. 2017). A recent survey conducted by Villarroel et al. (2016) found that from

68 responding aquaponic actors spread across 21 European countries, 75% were involved in research activities and 30.8% in production, with only 11.8% of those surveyed actually selling fish or plants in the past 12 months. It is clear that the field of aquaponics in Europe is still mainly shaped by actors from research. In this developmental environment, we believe the next phase of aquaponic research will be crucial to developing the future sustainability and food security potential of this technology.

Interviews (König et al. 2018) and the quantitative surveys (Villarroel et al. 2016) of the European aquaponic field have indicated there is mixed opinion regarding the vision, motivations and expectations about the future of aquaponics. In light of this, Konig et al. (2018) have raised concerns that a diversity of visions for aquaponic technology might hinder the coordination between actors and ultimately disrupt the development of 'a realistic corridor of acceptable development paths' for the technology (König et al. 2018). From an innovation systems perspective, emergent innovations that display an unorganised diversity of visions can suffer from 'directionality failure' (Weber and Rohracher 2012) and ultimately fall short of their perceived potentials. Such perspectives run in line with positions from sustainability science that stress the importance of 'visions' for creating and pursuing desirable futures (Brewer 2007). In light of this, we offer up one such vision for the field of aquaponics. We argue that aquaponics research must refocus on a radical sustainability and food security agenda that is fit for the impending challenges faced in the Anthropocene.

Towards a 'Sustainability First' Paradigm

As we saw earlier, it has been stressed that the goal to move towards sustainable intensification grows from the acknowledgment of the limits of the conventional agricultural development paradigm and its systems of innovation. Acknowledging the need for food system innovations that exceed the traditional paradigm and that can account for the complexity arising from sustainability and food security issues, Fischer et al. (2007) have called for no less than 'a new model of sustainability' altogether. Similarly, in their recent plea for global efforts towards sustainable intensification, Rockström et al. (2017) have pointed out that a paradigm shift in our food system entails challenging the dominant research and development patterns that maintain the 'productivity first' focus whilst subordinating sustainability agendas to a secondary, 'mitigating' role. Instead, they call for a reversal of this paradigm so that 'sustainable principles become the *entry point* for generating productivity enhancements'. Following this, we suggest a *sustainability first* vision for aquaponics as one possible orientation that can both offer coherence to the field and guide its development towards the proclaimed goals of sustainability and food security.

As with most calls for sustainability, our *sustainability first* proposal might sound rather obvious and unchallenging at first glance, if not completely redundant—surely, we could say, aquaponics is all about sustainability. But history would remind us that making sustainability claims is an agreeable task, whereas securing sustainability outcomes is far less certain (Keil 2007). As we have argued, the 'sustainability' of aquaponics currently exists as potential. Just how this potential translates into sustainability outcomes must be a concern for our research community.

Our 'sustainability first' proposal is far from straightforward. First and foremost, this proposal demands that, if our field is to justify itself on the grounds of sustainability, we must get to grips with the nature of sustainability itself. In this regard, we feel there is much to be learned from the growing arena of sustainability science as well as Science and Technology Studies (STS). We will find that maintaining a sustainability focus within aquaponic research represents a potentially huge shift in the direction, composition and ambition of our research community. Such a task is necessary if we are to direct the field towards coherent and realistic goals that remain focussed on sustainability and food security outcomes that are relevant for the Anthropocene.

Taking sustainability seriously is a massive challenge. This is because, at its core, sustainability is fundamentally an *ethical* concept raising questions about the value of nature, social justice, responsibilities to future generations, etc. and encompasses the multidimensional character of human-environment problems (Norton 2005). As we discussed earlier, the sustainability thresholds that might be drawn up concerning agricultural practices are diverse and often cannot be reconciled in entirety, obligating the need for 'trade-offs' (Funtowicz and Ravetz 1995). Choices have to be made in the face of these trade-offs and most often the criteria upon which such choices are based depend not only upon scientific, technical or practical concerns but also on norms and moral values. It goes without saying, there is little consensus on how to make these choices nor is there greater consensus on the norms and moral values themselves. Regardless of this fact, inquiries into values are largely absent from the mainstream sustainability science agenda, yet as Miller et al. (2014) assert, 'unless the values [of sustainability] are understood and articulated, the unavoidable political dimensions of sustainability will remain hidden behind scientific assertions'. Such situations prevent the coming together of and democratic deliberation between communities—a certain task for achieving more sustainable pathways.

Taking note of the prominent place of values in collective action towards sustainability and food security, scholars from the field of science and technology studies have highlighted that rather than be treated as an important externality to research processes (often dealt with separately or after the fact), values must be moved upstream in research agendas (Jasanoff 2007). When values become a central part of sustainability research, along comes the acknowledgement that decisions can no longer be based on technical criteria alone. This has potentially huge impacts on the research process, because traditionally what might have been regarded as the sole remit of 'expert knowledge' must now be opened up to other knowledge streams (for instance, 'lay', indigenous and practitioner knowledge) with all the epistemological difficulty this entails (Lawrence 2015). In response to these problems, sustainability science has emerged as a field that aims to transcend disciplinary boundaries and seeks to involve non-scientists in solution-oriented, context-determined, research processes that are focused on outcome generation (Miller et al. 2014).

A key question in these discussions is knowledge. Sustainability problems are often caused by the complex interplay of diverse social–ecological factors, and the knowledge needed for effectively governing these challenges has become progressively more dispersed and specialised (Ansell and Gash 2008). The knowledge required for understanding how sustainability concerns hang together is too complex to be organised by a single body and results in the need to integrate different types of knowledge in new ways. This is certainly the case for our own field: like other modes of sustainable intensification (Caron et al. 2014), aquaponic systems are characterised by inherent complexity (Junge et al. 2017) which places great emphasis on new forms of knowledge production (FAO 2013). Complexity of aquaponic systems derives not only from their 'integrated' character but stems also from the wider economic, institutional and political structures that impact the delivery of aquaponics and its sustainability potential (König et al. 2016). Developing solutions towards sustainable aquaponic food systems may well involve contending with diverse realms of understanding from engineering, horticultural, aquacultural, microbiological, ecological, economic and public health research, to the practical and experiential knowledge concerns of practitioners, retailers and consumers. What this amounts to is not just a grouping together of ideas and positions, but entails developing entirely novel modes of knowledge production and an appreciation to bridge 'knowledge gaps' (Caron et al. 2014). Abson et al. (2017) have identified three key requirements of new forms of knowledge production that can foster sustainability transformations: (i) the explicit inclusion of values, norms and context characteristics into the research process to produce 'socially robust' knowledge; (ii) mutual learning processes between science and society, involving a rethink of the role of science in society; and (iii) a problem- and solution-oriented research agenda. Drawing upon these three insights can help our field develop what we call a 'critical sustainability knowledge' for aquaponics. Below we discuss three areas our research community can address that we consider crucial to unlocking the sustainability potential of aquaponics: partiality, context and concern. Developing an understanding of each of these points will help our field pursue a solution-oriented approach for aquaponic sustainability and food security outcomes.

'Critical Sustainability Knowledge' for Aquaponics

Partiality

Despite contemporary accounts of sustainability that underline its complex, multidimensional and contested character, in practice, much of the science that engages with sustainability issues remains fixed to traditional, disciplinary perspectives and actions (Miller et al. 2014). Disciplinary knowledge, it must be said, has obvious value and has delivered huge advances in understanding since antiquity. Nevertheless, the appreciation and application of sustainability issues through traditional disciplinary channels has been characterised by the historic failure to facilitate the deeper societal change needed for issues such as the one we contend with here—the sustainable transformation of the food system paradigm (Fischer et al. 2007).

The articulation of sustainability problems through traditional disciplinary channels often leads to 'atomised' conceptualisations that view biophysical, social and economic dimensions of sustainability as compartmentalised entities and assume these can be tackled in isolation (e.g. Loos et al. 2014). Instead of viewing sustainability issues as a convergence of interacting components that must be addressed together, disciplinary perspectives often promote 'techno-fixes' to address what are often complex multidimensional problems (e.g. Campeanu and Fazey 2014). A common feature of such framings is that they often imply that sustainability problems can be resolved without consideration of the structures, goals and values that underpin complex problems at deeper levels, typically giving little consideration to the ambiguities of human action, institutional dynamics and more nuanced conceptions of power.

The practice of breaking a problem down into discrete components, analysing these in isolation and then reconstructing a system from interpretations of the parts has been a hugely powerful methodological insight that traces its history back to the dawn of modernity with the arrival of Cartesian reductionism (Merchant 1981). Being a key tenet of the production of objective knowledge, this practice forms the bedrock of most disciplinary effort in the natural sciences. The importance of objective knowledge, of course, is in that it provides the research community with 'facts'; precise and reproducible insights about generally dispersed phenomena. The production of facts was the engine room of innovation that propelled the Green Revolution. Science fuelled 'expert knowledge' and provided penetrating information about dynamics in our food production systems that remained invariant through change in time, space or social location. Building a catalogue of this kind of knowledge, and deploying it as what Latour (1986) calls 'immutable mobiles', formed the basis of the universal systems of monocropping, fertilisation and pest control that characterise the modern food system (Latour 1986).

But this form of knowledge production has weaknesses. As any scientist knows, in order to gain significant insights, this method must be strictly applied. It has been shown that this knowledge production is 'biased toward those elements of nature which yield to its method and toward the selection of problems most tractable to solutions with the knowledge thereby produced' (Kloppenburg 1991). A clear example of this would be our imbalanced food security research agenda that heavily privileges production over conservation, sustainability or food sovereignty issues (Hunter et al. 2017). Most high-profile work on food security concentrates on production (Foley et al. 2011), emphasising material flows and budgets over deeper issues such as the structures, rules and values that shape food systems. The simple fact is that because we know more about material interventions it is easier to design, model and experiment on these aspects of the food system. As Abson et al. (2017: 2) point out: 'Much scientific lead sustainability applications assume some of the most challenging drivers of unsustainability can be viewed as "fixed system properties" that can be addressed in isolation'. In pursuing the paths along which experimental success is most often realised, 'atomised' disciplinary approaches neglect those areas where other approaches might prove rewarding. Such epistemological 'blind spots' mean that sustainability interventions are often geared towards highly tangible aspects that may be simple to envisage and implement, yet have weak potential for 'leveraging' sustainable transition or deeper system change (Abson et al. 2017). Getting to grips with the limits and partialities of our disciplinary knowledge is one aspect that we stress when we claim the need to develop a 'critical sustainability knowledge' for aquaponics.

Viewed from disciplinary perspectives the sustainability credentials of aquaponic systems can be more or less simple to define (for instance, water consumption, efficiency of nutrient recycling, comparative yields, consumption of non-renewable inputs, etc.). Indeed, the more narrowly we define the sustainability criteria, the more straightforward it is to test such parameters, and the easier it is to stamp the claim of sustainability on our systems. The problem is that we can engineer our way to a form of sustainability that only few might regard as sustainable. To paraphrase Kläy et al. (2015), when we transform our original concern of how to realise a sustainable food system into a 'matter of facts' (Latour 2004) and limit our research effort to the analysis of these facts, we subtly but profoundly change the problem and direction of research. Such an issue was identified by Churchman (1979:4–5) who found that because science addresses mainly the identification and the solution of problems, and not the systemic and related ethical aspects, there is always the risk that the solutions offered up may even increase the unsustainability of development—what he called the 'environmental fallacy' (Churchman 1979).

We might raise related concerns for our own field. Early research in aquaponics attempted to answer questions concerning the environmental potential of the technology, for instance, regarding water discharge, resource inputs and nutrient recycling, with research designed around small-scale aquaponic systems. Although admittedly narrow in its focus, this research generally held sustainability concerns in focus. Recently, however, we have detected a change in research focus. This is raised in Chapter of this book, whose authors share our own view, observing that research 'in recent years has increasingly shifted towards economic feasibility in order to make aquaponics more productive for large-scale farming applications'. Discussions, we have found, are increasingly concerned with avenues of efficiency and profitability that often fix the potential of aquaponics against its perceived competition with other large-scale production methods (hydroponics and RAS). The argument appears to be that only when issues of system productivity are solved, through efficiency measures and technical solutions such as optimising growth conditions of plants and fish, aquaponics becomes economically competitive with other industrial food production technologies and is legitimated as a food production method.

We would certainly agree that economic viability is an important constituent of the long-term resilience and sustainability potential of aquaponics. However, we would caution against too narrowly defining our research ethic—and indeed, the future vision of aquaponics—based on principles of production and profit alone. We worry that when aquaponic research is limited to efficiency, productivity and market competitiveness, the old logics of the Green Revolution are repeated and our claims to food security and sustainability become shallow. As we saw earlier, productionism has been understood as a process in which a logic of production overdetermines other activities of value within agricultural systems (Lilley and Papadopoulos 2014). Since sustainability inherently involves a complex diversity of values, these narrow avenues of research, we fear, risk the articulation of aquaponics within a curtailed vision of sustainability. Asking the question 'under what circumstances can aquaponics outcompete traditional large-scale food production methods?' is not the same as asking 'to what extent can aquaponics meet the sustainability and food security demands of the Anthropocene?'.

Context

Knowledge production through traditional disciplinary pathways involves a loss of context that can narrow our response to complex sustainability issues. The multidimensional nature of food security implies that 'a single globally valid pathway to sustainable intensification does not exist' (Struik and Kuyper 2014). The physical, ecological and human demands placed on our food systems are context-bound and, as such, so are the sustainability and food security pressures which flow from these needs. Intensification requires contextualisation (Tittonell and Giller 2013). Sustainability and food security are outcomes of 'situated' practices, and cannot be extracted from the idiosyncrasies of context and 'place' that are increasingly seen as important factors in the outcomes of such (Altieri 1998; Hinrichs 2003; Reynolds et al. 2014). Added to this, the Anthropocene throws up an added task: localised forms of knowledge must be coupled with 'global' knowledge to produce sustainable solutions. The Anthropocene problematic places a strong need upon us to recognise the interconnectedness of the world food system and our globalised place within it: The particular way sustainable intensification is achieved in one part of the planet is likely to have ramifications elsewhere (Garnett et al. 2013). Developing a 'critical sustainability knowledge' means opening up to the diverse potentials and restraints that flow from contextualised sustainability concerns.

One of the main ruptures proposed by ecological intensification is the movement away from the chemical regulation that marked the driving force of agricultural development during the industrial revolution and towards biological regulation. Such a move reinforces the importance of local contexts and specificities. Although dealing most often with traditional, small-holder farming practices, agroecological methods have shown how context can be attended to, understood, protected and celebrated in its own right (Gliessman 2014). Studies of 'real' ecosystems in all their contextual complexity may lead to a 'feeling for the ecosystem'—critical to the pursuit of understanding and managing food production processes (Carpenter 1996).

The relevance of agroecological ideas need not be restricted to 'the farm'; the nature of closed-loop aquaponics systems demands a 'balancing' of co-dependent ecological agents (fish, plants, microbiome) within the limits and affordances of each particular system. Although the microbiome of aquaponics systems has only just begun to be analysed (Schmautz et al. 2017), complexity and dynamism is expected to exceed Recirculating Aquaculture Systems, whose microbiology is known to be affected by feed type and feeding regime, management routines, fish-associated microflora, make-up water parameters and selection pressure in the biofilters (Blancheton et al. 2013). What might be regarded as 'simple' in comparison to other farming methods, the ecosystem of aquaponics systems is nevertheless dynamic and requires care. Developing an 'ecology of place', where context is intentionality and carefully engaged with, can serve as a creative force in research, including scientific understanding (Thrift 1999; Beatley and Manning 1997).

The biophysical and ecological dynamics of aquaponic systems are central to the whole conception of aquaponics, but sustainability and food security potentials do not derive solely from these parameters. As König et al. (2016) point out, for aquaponic systems: 'different settings potentially affects the delivery of all aspects of sustainability: economic, environmental and social' (König et al. 2016). The huge configurational potential of aquaponics—from miniature to hectares, extensive to intensive, basic to high-tech systems—is quite atypical across food production technologies (Rakocy et al. 2006). The integrative character and physical plasticity of aquaponic systems means that the technology can be deployed in a wide variety of applications. This, we feel, is precisely the strength of aquaponic technology. Given the diverse and heterogeneous nature of sustainability and food security concerns in the Anthropocene, the great adaptability, or even 'hackability' (Delfanti 2013), of aquaponics offers much potential for developing 'custom-fit' food production (Reynolds et al. 2014) that is explicitly tailored to the environmental, cultural and nutritional demands of place. Aquaponic systems promise avenues of food production that might be targeted towards local resource and waste assimilative limits, material and technological availability, market and labour demands. It is for this reason that the pursuit of sustainability outcomes may well involve different technological developmental paths dependent upon locale (Coudel et al. 2013). This is a point that is beginning to receive increasing acknowledgement, with some commentators claiming that the urgency of global sustainability and food security issues in the Anthropocene demand an open and multidimensional approach to technological innovation. For instance, Foley et al. (2011:5) state: 'The search for agricultural solutions should remain technology neutral. There are multiple paths to improving the production, food security and environmental performance of agriculture, and we should not be locked into a single approach a priori, whether it be conventional agriculture, genetic modification or organic farming' (5) (Foley et al. 2011). We would highlight this point for aquaponics, as König et al. (2018: 241) have already done: 'there are several sustainability problems which aquaponics could address, but which may be impossible to deliver in one system setup. Therefore, future pathways will always need to involve a diversity of approaches'.

But the adaptability of aquaponics might be seen as a double-edged sword. Inspiration for specific 'tailor-made' sustainability solutions brings with it the difficulty of generalising aquaponic knowledge for larger-scale and repeatable purposes. Successful aquaponics systems respond to local specificities in climate, market, knowledge, resources, etc. (Villarroel et al. 2016; Love et al. 2015; Laidlaw and Magee 2016), but this means that changes at scale cannot easily proceed from the fractal replication of non-reproducible local success stories. Taking similar issues as these into account, other branches of ecological intensification research have suggested that the expression 'scaling up' must be questioned (Caron et al. 2014). Instead, ecological intensification is beginning to be viewed as a transition of multi-scalar processes, all of which follow biological, ecological, managerial and political 'own rules', and generate unique trade-off needs (Gunderson 2001).

Understanding and intervening in complex systems like this presents huge challenges to our research, which is geared towards the production of 'expert knowledge', often crafted in the lab and insulated from wider structures. The complex problem of food security is fraught with uncertainties that cannot be adequately resolved by resorting to the puzzle-solving exercises of Kuhnian 'normal science' (Funtowicz and Ravetz 1995). The necessity to account for 'specificity' and 'generality' in complex sustainably issues produces great methodological, organisational and institutional difficulties. The feeling is that to meet contextualised sustainability and food security goals, 'universal' knowledge must be connected to 'place-based' knowledge (Funtowicz and Ravetz 1995). For Caron et al. (2014), this means that 'scientists learn to continually go back and forth...' between these two dimensions, '. . .both to formulate their research question and capitalize their results. . . Confrontation and hybridization between heterogeneous sources of knowledge is thus essential' (Caron et al. 2014). Research must be opened up to wider circles of stakeholders and their knowledge streams.

Given the huge challenge on all accounts that such a scheme entails, a tempting resolution might be found in the development of more advanced 'environment-controlled' aquaponic farming techniques. Such systems work by cutting out external influences in production, maximising efficiency by minimising the influence of suboptimal, location-specific variables (Davis 1985). But we question this approach on a number of accounts. Given that the impulse of such systems lies in buffering food production from 'localised inconsistencies', there is always a risk that the localised sustainability and food security needs might also be externalised from system design and management. Cutting out localised anomalies in the search of the 'perfect system' must certainly offer tantalising efficiency potentials on paper, but we fear this type of problem-solving bypasses the specificity-generality problematic of sustainability issues in the Anthropocene without confronting them. Rather than a remedy, the result may well be an extension of the dislocated, 'one size fits all' approach to food production that marked the Green Revolution.

Current aquaponics research that follows either of the informal schools of 'decoupling' or 'closing the cycle' might well be an example of such framings. By pushing the productivity limits of either production side—aquaculture or hydroculture—inherent operational compromises of the ecological aquaponic principle become more apparent and become viewed as barriers to productivity that must be overcome. Framing the aquaponic problem like this results in solutions that involve more technology: patented one-way valves, condensation traps, high-tech oxygenators, LED lighting, additional nutrient dispensers, nutrient concentrators and so on. These directions repeat the knowledge dynamic of modern industrial agriculture that overly concentrated the expertise and power of food production systems into the hands of applied scientists engaged in the development of inputs, equipment and remote system management. We are unsure of how such technocratic measures might fit within a research ethic that places sustainability first. This is not an argument against high-tech, closed environment systems; we simply hope to emphasise that within a *sustainability first* paradigm, our food production technologies must be justified on the grounds of generating context-specific sustainability and food security outcomes.

Understanding that sustainability cannot be removed from the complexities of context or the potentials of place is to acknowledge that 'expert knowledge' alone cannot be held as guarantor of sustainable outcomes. This strikes a challenge to modes of centralised knowledge production based on experiments under controlled conditions and the way science might contribute to the innovation processes (Bäckstrand 2003). Crucial here is the design of methodological systems that ensure both the robustness and genericity of scientific knowledge is maintained along with its relevance to local conditions. Moving to conceptions like this requires a huge shift in our current knowledge production schemes and not only implies better integration of agronomic with human and political sciences but suggests a path of knowledge co-production that goes well beyond 'interdisciplinarity' (Lawrence 2015).

Here it is important to stress Bäckstrand's (2003: 24) point that the incorporation of lay and practical knowledge in scientific processes 'does not rest on the assumption that lay knowledge is necessarily "truer", "better" or "greener"'. Rather, as Leach et al. (2012: 4) point out, it stems from the idea that 'nurturing more diverse approaches and forms of innovation (social as well as technological) allows us to respond to uncertainty and surprise arising from complex, interacting biophysical and socioeconomic shocks and stresses'. Faced with the uncertainty of future environmental outcomes in the Anthropocene, a multiplicity of perspectives can prevent the narrowing of alternatives. In this regard, the potential wealth of experimentation occurring in 'backyard' and community projects across Europe represents an untapped resource which has until now received little attention from research circles. 'The small-scale sector...' Konig et al. (2018: 241) observe, '...shows optimism and a surprising degree of self-organization over the internet. There might be room for creating additional social innovations'. Given the multidimensional nature of issues in the Anthropocene, grassroots innovations, like the backyard aquaponics sector, draw from local knowledge and experience and work towards social and organisational forms of innovation that are, in the eyes of Leach et al. (2012: 4), 'at least as crucial as advanced science and technology'. Linking with community aquaponics groups potentially offers access to vibrant local food groups, local government and local consumers who are often enthusiastic about the prospects of collaborating with researchers. It is worth noting that in an increasingly competitive funding climate, local communities offer a well of resources—intellectual, physical and monetary—that often get overlooked but which can supplement more traditional research funding streams (Reynolds et al. 2014).

As we know, currently, large-scale commercial projects face high marketing risks, strict financing deadlines, as well as high technological and management complexity that makes collaboration with outside research organisations difficult. Because of this, we would agree with König et al. (2018) who find advantages for experimentation with smaller systems that have reduced complexity and are tied down by fewer legal regulations. The field must push to integrate these organisations within participatory, citizen-science research frameworks, allowing academic research to more thoroughly mesh with forms of aquaponics working in the world. In the absence of formalised sustainability measures and protocols, aquaponic enterprises risk legitimation issues when their produce is marketed on claims of sustainability. One clear possibility of participatory research collaborations would be the joint production of much needed 'situation-specific sustainability goals' for facilities that could form the 'basis for system design' and bring 'a clear marketing strategy' (König et al. 2018). Working towards outcomes like these might also improve the transparency, legitimacy and relevance of our research endeavours (Bäckstrand 2003).

The European research funding climate has begun to acknowledge the need to shift research orientation by including the requirement in recent project funding calls of implementing the so-called 'living labs' into research projects (Robles et al. 2015). Starting in June 2018, the Horizon 2020 project proGIreg (H2020-SCC-2016-2017) is going to include a living lab for the exemplary implementation of the so-called nature-based systems (NBS), one of which will be a community designed, community-built and community-operated aquaponic system in a passive solar greenhouse. The project, with 36 partners in 6 countries, aims to find innovative ways to productively utilise green infrastructure of urban and peri-urban environments, building upon the co-production concepts developed in its currently running sibling project, CoProGrün.

The researchers' working packages regarding the aquaponic part of the project are going to be threefold. One part will be about raising the so-called technology readiness level (TRL) of aquaponics, a research task without explicit collaboration with laypersons and the community. Resource utilisation of current aquaponic concepts and resource optimisation potential of additional technical measures are the core objectives of this task. While at first glance this task seems to follow the above-criticised paradigm of productivity and yield increase, evaluation criteria for different measures will include more multifaceted aspects such as ease of implementation, understandability, appropriateness and transferability. A second focus will be support of the community planning, building and operational processes, which seeks to integrate objective knowledge and practitioner knowledge generation. A meta-objective of this process will be the observation and the moderation of the relevant community collaboration and communication processes. In this approach, moderation is actively expected to alter observation, illustrating a deviation from the traditional research routines of fact building and repeatability. A third package encompasses research on political, administrative, technical and financial obstacles. The intention here is to involve a wider collection of stakeholders, from politicians and decision-makers to planners, operators and neighbours, with research structures developed to bring together each of these specific perspectives. Hopefully, this more holistic method opens a path to the 'sustainability first' approach proposed in this chapter.

Concern

Recognising aquaponics as a multifunctional form of food production faces large challenges. As has been discussed, grasping the notion of 'multifunctional agriculture' is more than just a critical debate on what constitutes 'post-productionism' (Wilson 2001); this is because it seeks to move understandings of our food system to positions that better encapsulates the diversity, nonlinearity and spatial heterogeneity that are acknowledged as key ingredients to a sustainable and just food system. It is important to remember that the very notion of 'multifunctionality' in agriculture arose during the 1990s as 'a consequence of the undesired and largely unforeseen environmental and societal consequences and the limited cost-effectiveness of the European Common Agricultural Policy (CAP), which mainly sought to boost agrarian outputs and the productivity of agriculture' (270) (Cairol et al. 2009). Understanding that our political climates and institutional structures have been unconducive to sustainable change is a point we must not forget. As others have pointed out in adjacent agronomic fields, understanding and unlocking the richness of food production contributions to human welfare and environmental health will necessarily involve a *critical* dimension (Jahn 2013). This insight, we feel, must feature more strongly in aquaponics research.

We chose the word 'concern' here carefully. The word concern carries different connotations to 'critique'. Concern carries notions of anxiety, worry and trouble. Anxiety comes when something disrupts what could be a more healthy or happy or secure existence. It reminds us that to do research in the Anthropocene is to acknowledge our drastically unsettling place in the world. That our 'solutions' always carry the possibility of trouble, whether this be ethical, political or environmental. But concern has more than just negative connotations. To concern also means to 'be about', to 'relate to' and also 'to care'. It reminds us to question what our research is about. How our disciplinary concerns relate to other disciplines as well as wider issues. Crucially, sustainability and food security outcomes require us to care about the concerns of others.

Considerations such as these make up a third aspect of what we mean when we call for a *'Critical sustainability knowledge'* for aquaponics. As a research community, it is crucial that we develop an understanding of the structural factors which impinge upon and restrict the effective social, political and technological innovation of aquaponics. Technical change relies upon infrastructure, financing capacities, market organisations as well as labour and land rights conditions (Röling 2009). When the role of this wider framing is assumed only as an 'enabling environment', often the result is that such considerations are left outside of the research effort. This is a point which serves to easily justify the failure of technology-based, top-down development drives (Caron 2000). In this regard, the techno-optimistic discourse of contemporary aquaponics, in its failure to apprehend wider structural resistance to the development of sustainable innovation, would serve as a case example.

As an important potential form of sustainable intensification, aquaponics needs to be recognised as being embedded in and linked to different social, economic and organisational forms at various scales potentially from household, value chain, food system and beyond including also other political levels. Thankfully, moves towards attending to the wider structural difficulties that aquaponic technology faces have recently been made, with König et al. (2018) offering a view of aquaponics through an 'emerging technological innovation system' lens. König et al. (2018) have shown how the challenges to aquaponics development derive from: (1) system complexity, (2) the institutional setting and (3) the sustainability paradigm it attempts to impact. The aquaponic research field needs to respond to this diagnosis.

The slow uptake and high chance of failure that aquaponics technology currently exhibits is an expression of the wider societal resistance that makes sustainable innovation such a challenge, as well as our inability to effectively organise against such forces. As König et al. (2018) note, the high-risk environment that currently exists for aquaponic entrepreneurs and investors forces startup facilities across Europe to focus on production, marketing and market formation, over the delivery of sustainability credentials. Along these lines, Alkemade and Suurs (2012) remind us, 'market forces alone cannot be relied upon to realize desired sustainability transitions'; rather, they point out, insight into the dynamics of innovation processes is needed if technological change can be guided along more sustainable trajectories (Alkemade and Suurs 2012).

The difficulties aquaponic businesses face in Europe suggest the field currently lacks the necessary market conditions, with 'consumer acceptance'—an important factor enabling the success of novel food system technologies—acknowledged as a possible problem area. From this diagnosis, there has been raised the problem of 'consumer education' (Miličić et al. 2017). Along with this, we would stress that collective education is a key concern for questions of food system sustainability. But accounts like these come with risks. It is easy to fall back on traditional modernist conceptions regarding the role of science in society, assuming that 'if only the public understood the facts' about our technology they would choose aquaponics over other food production methods. Accounts like these assume too much, both about the needs of 'consumers', as well as the value and universal applicability of expert knowledge and technological innovation. There is a need to seek finer-grain and more nuanced accounts of the struggle for sustainable futures that move beyond the dynamic of consumption (Gunderson 2014) and have greater sensitivity to the diverse barriers communities face in accessing food security and implementing sustainable action (Carolan 2016; Wall 2007).

Gaining insight into innovation processes puts great emphasis upon our knowledge-generating institutions. As we have discussed above, sustainability issues demand that science opens up to public and private participatory approaches entailing knowledge co-production. But in terms of this point, it's worth noting that huge challenges lay in store. As Jasanoff (2007: 33) puts it: 'Even when scientists recognize the limits of their own inquiries, as they often do, the policy world, implicitly encouraged by scientists, asks for more research'. The widely held assumption that *more* objective knowledge is the key to bolstering action towards sustainability runs contrary to the findings of sustainability science. Sustainability outcomes are actually more closely tied deliberative knowledge processes: building greater awareness of the ways in which experts and practitioners frame sustainability issues; the values that are included as well as excluded; as well as effective ways of facilitating communication of diverse knowledge and dealing with conflict if and when it arises (Smith and Stirling 2007; Healey 2006; Miller and Neff 2013; Wiek et al. 2012). As Miller et al. (2014) point out, the continuing dependence upon objective knowledge to adjudicate sustainability issues represents the persistence of the modernist belief in rationality and progress that underwrites almost all knowledge-generating institutions (Horkheimer and Adorno 2002; Marcuse 2013).

It is here where developing a critical sustainability knowledge for aquaponics shifts our attention to our own research environments. Our increasingly 'neoliberalised' research institutions exhibit a worrying trend: the rollback of public funding for universities, the increasing pressure to get short-term results, the separation of research and teaching missions, the dissolution of the scientific author, the contraction of research agendas to focus on the needs of commercial actors, an increasing reliance on market take-up to adjudicate intellectual disputes and the intense fortification of intellectual property in the drive to commercialise knowledge, all of which have been shown to impact on the production and dissemination of our research, and indeed all are factors that impact the nature of our science (Lave et al. 2010). One question that must be confronted is whether our current research environments are fit for the examination of complex sustainability and long-term food security targets that must be part of aquaponic research. This is the key point we would like to stress—if sustainability is an outcome of multidimensional collective deliberation and action, our own research endeavours, thoroughly part of the process, must be viewed as something that can be innovated towards sustainability outcomes also. The above-mentioned Horizon 2020 project proGIreg may be an example of some ambitious first steps towards crafting new research environments, but we must work hard to keep the research process itself from slipping out of view. Questions might be raised about how these potentially revolutionary measures of 'living labs' might be implemented from within traditional funding logics. For instance, calls for participatory approaches foreground the conceptual importance of open-ended outcomes, while at the same time requiring the intended spending of such living labs to be predefined. Finding productive ways out of traditional institutional barriers is an ever-present concern.

Our modern research environments can no longer be regarded as having a privileged isolation from the wider issues of society. More than ever our innovation-driven biosciences are implicated in the agrarian concerns of the Anthropocene (Braun and Whatmore 2010). The field of Science and Technology Studies teaches us that technoscientific innovations come with serious ethico-political implication. A 30-year-long discussion in this field has moved well beyond the idea that technologies are simply 'used' or 'misused' by different socio-political interests after the hardware has been 'stabilised' or legitimated through objective experimentation in neutral lab spaces (Latour 1987; Pickering 1992). The 'constructivist' insight in STS analyses goes beyond the identification of politics inside labs (Law and Williams 1982; Latour and Woolgar 1986 [1979]) to show that the technologies we produce are not 'neutral' objects but are in fact infused with 'world-making' capacities and political consequence.

The aquaponics systems we help to innovate are filled with future making capacity, but the consequences of technological innovation are seldom a focus of study. To paraphrase Winner (1993), what the introduction of new artefacts means for people's sense of self, for the texture of human/nonhuman communities, for qualities of everyday living within the dynamic of sustainability and for the broader distribution of power in society, these have not traditionally been matters of explicit concern. When classic studies (Winner 1986) ask the question 'Do artefacts have politics?', this is not only a call to produce more accurate examinations of technology by including politics in accounts of the networks of users and stakeholders, though this is certainly needed; it also concerns us researchers, our modes of thought and ethos that affect the politics (or not) we attribute to our objects (de la Bellacasa 2011; Arboleda 2016). Feminist scholars have highlighted how power relations are inscribed into the very fabric of modern scientific knowledge and its technologies. Against alienated and abstract forms of knowledge, they have innovated key theoretical and methodological approaches that seek to bring together objective and subjective views of the world and to theorise about technology from the starting point of practice (Haraway 1997; Harding 2004). Aware of these points, Jasanoff (2007) calls for the development of what she calls 'technologies of humility': 'Humility instructs us to think harder about how to reframe problems so that their ethical dimensions are brought to light, which new facts to seek and when to resist asking science for clarification. Humility directs us to alleviate known causes of people's vulnerability to harm, to pay attention to the distribution of risks and benefits, and to reflect on the social factors that promote or discourage learning'.

An important first step for our field to take towards understanding better the political potentials of our technology would be to encourage the expansion of the field out into critical research areas that are currently underrepresented. Across the Atlantic in the US and Canada similar moves like this have already been made, where an interdisciplinary approach has progressively developed into the critical field of political ecology (Allen 1993). Such projects not only aim to combine agriculture and land use patterns with technology and ecology, but furthermore, also emphasise the integration of socioeconomic and political factors (Caron et al. 2014). The aquaponics research community in America has begun to acknowledge the expanding resources of food sovereignty research, exploring how urban communities can be re-engaged with the principles of sustainability, whilst taking more control over their food production and distribution (Laidlaw and Magee 2016). Food sovereignty has become a huge topic that precisely seeks to intervene into food systems that are overdetermined by disempowering capitalist relations. From food sovereignty perspectives, the corporate control of the food system and the commodification of food are seen as predominant threats to food security and the natural environment (Nally 2011). We would follow Laidlaw and Magee's (2016) view that community-based aquaponics enterprises 'represent a new model for how to blend local agency with scientific innovation to deliver food sovereignty in cities'.

Developing a *'critical sustainability knowledge'* for aquaponics means resisting the view that society and its institutions are simply neutral domains that facilitate the linear progression towards sustainable innovation. Many branches of the social sciences have contributed towards an image of society that is infused with asymmetric power relations, a site of contestation and struggle. One such struggle concerns the very meaning and nature of sustainability. Critical viewpoints from wider fields would underline that aquaponics is a technology ripe with both political potential and limitation. If we are serious about the sustainability and food security credentials of aquaponics, it becomes crucial that we examine more thoroughly how our expectations of this technology relate to on-the-ground experience, and in turn, find ways of integrating this back into research processes. We follow Leach et al. (2012) here who insist on the need for finer-grained considerations regarding the performance of sustainable innovations. Apart from the claims, just who or what stands to benefit from such interventions must take up a central place in the aquaponic innovation process. Lastly, as the authors of Chapter have made clear, the search for a lasting paradigm shift will require the ability to place our research into policy circuits that make legislative environments more conducive to aquaponics development and enable larger-scale change. Influencing policy requires an understanding of the power dynamics and political systems that both enable and undermine the shift to sustainable solutions.

Conclusion: Aquaponic Research into the Anthropocene

The social–biophysical pressures *of* and *on* our food system converge in the Anthropocene towards what becomes seen as an unprecedented task for the global community, requiring 'nothing less than a planetary food revolution' (Rockström et al. 2017). The Anthropocene requires food production innovations that exceed traditional paradigms, whilst at the same time are able to acknowledge the complexity arising from the sustainability and food security issues that mark our times. Aquaponics is one technological innovation that promises to contribute much towards these imperatives. But this emergent field is in an early stage that is characterised by limited resources, market uncertainty, institutional resistance and high risks of failure—an innovation environment where hype prevails over demonstrated outcomes. The aquaponics research community potentially holds an important place in the development path of this technology. As an aquaponics research community, we need to craft viable visions for the future.

We propose one such vision when we call for a 'sustainability first' research programme. Our vision follows Rockström et al.'s (2017) diagnosis that paradigm change requires shifting the research ethic away from traditional productivist avenues so that sustainability becomes the central locus of the innovation process. This task is massive because the multidimensional and context-bound nature of sustainability and food security issues is such that they cannot be resolved solely through technical means. The ethical- and value-laden dimensions of sustainability require a commitment to confront the complexities, uncertainty, ignorance and contestation that ensue such issues. All this places great demands on the knowledge we produce; not only how we distribute and exchange it, but also its very nature.

We propose the aquaponic field needs to pursue a 'critical sustainability knowledge'. When König et al. (2018) ask what sustainability experimentation settings would be needed to enable science, business, policy and consumers to 'answer sustainability questions without repeating the development path of either [RAS or hydroponics]', the point is clear—we need to learn from the failures of the past. The current neoliberal climate is one that consistently opens 'sustainability' discussion up to (mis)appropriation as 'agribusiness mobilises its resources in an attempt to dominate discourse and to make its meaning of "alternative agriculture" the universal meaning' (Kloppenburg 1991). We need to build a critical sustainability knowledge that is wise to the limits of technocratic routes to sustainability, which is sensitive to the political potential of our technologies as well as the structural forms of resistance that limit their development.

A critical sustainability knowledge builds awareness of the limits of its own knowledge pathways and opens up to those other knowledge streams that are often pushed aside in attempts to expand scientific understanding and technological capacity. This is a call for interdisciplinarity and the depth it brings, but it goes further than this. Sustainability and food security outcomes have little impact if they can only be generated in the lab. Research must be contextualised: we need 'to produce and embed scientific knowledge into local innovation systems' (51) (Caron et al. 2014). Building co-productive links with aquaponics communities already existing in society means forging the social and institutional structures that can enable our communities to continually learn and adapt to new knowledge, values, technologies and environmental change. Together, we need to deliberate on the visions and the values of our communities and explore the potential sociotechnical pathways that might realise such visions. Central to this, we need systems of organising and testing the sustainability and food security claims that are made of this technology (Pearson et al. 2010; Nugent 1999) so that greater transparency and legitimation might be brought to the entire field: entrepreneurs, enterprises, researchers and activists alike.

If all this seems like a tall order, that's because it is. The Anthropocene calls for a huge rethink in the way society is being organised, and our food system is central to this. There is a chance, we believe, that aquaponics has a part to play in this. But if our hopes are not to get lost in the hype bubble of hollow sustainability chatter that marks our neoliberal times, we have to demonstrate that aquaponics offers something different. As a final remark, we revisit de la Bellacasa's (2015) point that: 'agricultural intensification is not only a quantitative orientation (yield increase), but entails a "way of life"'. If this is the case, then the pursuit of *sustainable intensification* demands that we find a new way of living. We need sustainability solutions that acknowledge this fact and research communities that are responsive to it.

[1] For instance, consider the following statement issued by Monsanto: 'The main uses of GM crops are to make them insecticide- and herbicide tolerant. They don't inherently increase the yield. They protect the yield'. Quoted in E. Ritch, 'Monsanto Strikes Back at Germany, UCS', Cleantech.com (April 17, 2009). Accessed on July 18, 2009.

[2] Especially important here are the effects of climate change, as well as the 'superweed' phenomenon of increasingly resistant pests that significantly diminish yields.

[3] Productivist discourse invariably ignores Amartya Sen's (1981, 154; Roberts 2008, 263; WFP 2009, 17) classic point that the volume and availability of food alone is not a sufficient explanation for the persistence of world hunger. It is well established that enough food exists to feed in excess of the world's current population (OECD 2009, 21)

[4] Although the calculations are complex and contested, one common estimate is that industrial agriculture requires an average 10 calories of fossil fuels to produce a single calorie of food (Manning 2004), which might rise to 40 calories in beef (Pimentel 1997).

[5] Externalities of our current food system are often ignored or heavily subsidised away. Moore (2015: 187) describes the situation as 'a kind of "ecosystem services" in reverse': 'Today, a billion pounds of pesticides and herbicides are used each year in American agriculture. The long recognized health impacts have been widely studied. Although the translation of such "externalities" into the register of accumulation is imprecise, their scale is impressive, totalling nearly $17 billion in unpaid costs for American agriculture in the early twenty-first century'. On externalities see: Tegtmeier and Duffy (2004).

References

Abson DJ, Fischer J, Leventon J, Newig J, Schomerus T, Vilsmaier U, von Wehrden H, Abernethy P, Ives CD, Jager NW (2017) Leverage points for sustainability transformation. Ambio 46:30–39

Alexandratos N, Bruinsma J (2012) World agriculture towards 2030/2050: the 2012 revision, ESA working paper. FAO, Rome

Alkemade F, Suurs RA (2012) Patterns of expectations for emerging sustainable technologies. Technol Forecast Soc Chang 79:448–456

Allen P (1993) Food for the future: conditions and contradictions of sustainability. Wiley, Chichester

Altieri MA (1998) Ecological impacts of industrial agriculture and the possibilities for truly sustainable farming. Mon Rev 50:60

Altieri MA (2007) Fatal harvest: old and new dimensions of the ecological tragedy of modern agriculture. In: Sustainable resource management. Edward Elgar, Londres, pp 189–213

Altvater E, Crist E, Haraway D, Hartley D, Parenti C, Mcbrien J (2016) Anthropocene or capitalocene?: nature, history, and the crisis of capitalism. PM Press, Oakland

Ansell C, Gash A (2008) Collaborative governance in theory and practice. J Public Adm Res Theory 18:543–571

Arboleda M (2016) Revitalizing science and technology studies: a Marxian critique of more-than-human geographies. Environt Plan D Soc Space 35:360–378. https://doi.org/10.1177/0263775816664099

Assessment ME (2003) Ecosystems and human well-being: biodiversity synthesis; a report of the millennium ecosystem assessment. Island Press, Washington, DC

Bäckstrand K (2003) Civic science for sustainability: reframing the role of experts, policy-makers and citizens in environmental governance. Glob Environ Politics 3:24–41

Battisti DS, Naylor RL (2009) Historical warnings of future food insecurity with unprecedented seasonal heat. Science 323:240–244

Baulcombe D, Crute I, Davies B, Dunwell J, Gale M, Jones J, Pretty J, Sutherland W, Toulmin C (2009) Reaping the benefits: science and the sustainable intensification of global agriculture. The Royal Society, London

Beatley T, Manning K (1997) The ecology of place: planning for environment, economy, and community. Island Press, Washington, DC

Blancheton J, Attramadal K, Michaud L, D'Orbcastel ER, Vadstein O (2013) Insight into bacterial population in aquaculture systems and its implication. Aquac Eng 53:30–39

Bommarco R, Kleijn D, Potts SG (2013) Ecological intensification: harnessing ecosystem services for food security. Trends Ecol Evol 28:230–238

Bradley A (2011) Originary technicity: the theory of technology from Marx to Derrida. Palgrave Macmillan Basingstoke, Hants

Braun B, Whatmore SJ (2010) Political matter: technoscience, democracy, and public life. University of Minnesota Press, Minneapolis

Brewer GD (2007) Inventing the future: scenarios, imagination, mastery and control. Sustain Sci 2:159–177

Bringezu S, Schütz H, Pengue W, O'Brien M, Garcia F, Sims R, Howarth RW, Kauppi L, Swilling M, Herrick J (2014) Assessing global land use: balancing consumption with sustainable supply. United Nations Environment Programme, Nairobi

Brooks S, Leach M, Millstone E, Lucas H (2009) Silver bullets, grand challenges and the new philanthropy. STEPS Centre, Brighton

Brussaard L, Caron P, Campbell B, Lipper L, Mainka S, Rabbinge R, Babin D, Pulleman M (2010) Reconciling biodiversity conservation and food security: scientific challenges for a new agriculture. Curr Opin Environ Sustain 2:34–42

Câmpeanu CN, Fazey I (2014) Adaptation and pathways of change and response: a case study from Eastern Europe. Glob Environ Chang 28:351–367

Cairol D, Coudel E, Knickel K, Caron P, Kröger M (2009) Multifunctionality of agriculture and rural areas as reflected in policies: the importance and relevance of the territorial view. J Environ Policy Plan 11:269–289

Cardinale BJ, Duffy JE, Gonzalez A, Hooper DU, Perrings C, Venail P, Narwani A, Mace GM, Tilman D, Wardle DA, Kinzig AP, Daily GC, Loreau M, Grace JB, Larigauderie A, Srivastava DS, Naeem S (2012) Biodiversity loss and its impact on humanity. Nature 486:59–67

Carolan M (2016) Adventurous food futures: knowing about alternatives is not enough, we need to feel them. Agric Hum Values 33:141–152

Caron P (2000) Decentralisation and multi-levels changes: challenges for agricultural research to support co-ordination between resource poor stakeholders and local governments

Caron P, Biénabe E, Hainzelin E (2014) Making transition towards ecological intensification of agriculture a reality: the gaps in and the role of scientific knowledge. Curr Opin Environ Sustain 8:44–52

Caron P, Reig E, Roep D, Hediger W, Cotty T, Barthelemy D, Hadynska A, Hadynski J, Oostindie H, Sabourin E (2008) Multifunctionality: refocusing a spreading, loose and fashionable concept for looking at sustainability? Int J Agric Resour Gov Ecol 7:301–318

Carpenter SR (1996) Microcosm experiments have limited relevance for community and ecosystem ecology. Ecology 77:677–680

Castree N (2015) Changing the Anthropo (s) cene geographers, global environmental change and the politics of knowledge. Dialogues Hum Geogr 5:301–316

Churchman CW (1979) The systems approach and its enemies. Basic Books, New York

Cohen T (2012) Anecographics: climate change and "Late" deconstruction. In: Impasses of the post-global: theory in the era of climate change, vol 2. Open Humanities Press, London, pp 32–57

Cohen T (2013) Telemorphosis: theory in the era of climate change, vol 1. Open Humanities Press

Collings DA (2014) Stolen future, broken present: the human significance of climate change. Open Humanities Press, London

Cooper M (2008) Life as surplus: biotechnology and capitalism in the neoliberal era. University of Washington Press, Seattle

COST (2013) Memorandum of understanding for the implementation of a European Concerted Research Action designated as COST Action FA1305: The EU Aquaponics Hub: realising sustainable integrated fish and vegetable production for the EU. European Cooperation in the field of Scientific and Technical Research – COST, Brussels

Cota GM (2011) Introduction: the posthuman life of agriculture. In: Cota GM (ed) Another technoscience is possible: agricultural lessons for the Posthumanities. Open Humanities Press, London

Coudel E, Devautour H, Soulard C-T, Faure G, Hubert B (2013) Renewing innovation systems in agriculture and food: how to go towards more sustainability? Wageningen Academic Publishers, Wageningen

Crutzen PJ, Stoermer EF (2000a) The "Anthropocene". Global Change Newsl 41:17–18. International Geosphere–Biosphere Programme (IGBP)

Crutzen PJ, Stoermer EF (2000b) Global change newsletter. Anthropocene 41:17–18

Davis N (1985) Controlled-environment agriculture-past, present and future. Food Technol 39:124–126

de La Bellacasa MP (2011) Matters of care in technoscience: assembling neglected things. Soc Stud Sci 41:85–106

de La Bellacasa MP (2015) Making time for soil: Technoscientific futurity and the pace of care. Soc Stud Sci 45:691–716

de Laet M, Mol A (2000) The Zimbabwe bush pump: mechanics of a fluid technology. Soc Stud Sci 30:225–263

Delaide B, Delhaye G, Dermience M, Gott J, Soyeurt H, Jijakli MH (2017) Plant and fish production performance, nutrient mass balances, energy and water use of the PAFF box, a small-scale aquaponic system. Aquac Eng 78:130

Delfanti A (2013) Biohackers. The politics of open science. Pluto Press, London

Desa U (2015) World population prospects: the 2015 revision, key findings and advance tables. Working Paper No Rep, No. ESA/P/WP. 241

Dobbs R, Oppenheim J, Thompson F, Brinkman M, Zornes M (2011) Resource revolution: meeting the world's energy, materials, food, and water needs. McKinsey & Company, London

Elzen,B, Janssen A, Bos A (2017) Portfolio of promises: designing and testing a new tool to stimulate transition towards sustainable agriculture. AgroEcological transitions. Wageningen University & Research

Evenson RE, Gollin D (2003) Assessing the impact of the green revolution, 1960 to 2000. Science 300:758–762

FAO (2013) Annotated bibliography on ecological intensification. Report of the Liberation EU Collaborative funded project (LInking farmland Biodiversity to Ecosystem seRvices for effective ecofunctional intensification)

Farmer B (1986) Perspectives on the 'Green Revolution'in South Asia. Mod Asian Stud 20 (1):175–199

Fischer J, Manning AD, Steffen W, Rose DB, Daniell K, Felton A, Garnett S, Gilna B, Heinsohn R, Lindenmayer DB (2007) Mind the sustainability gap. Trends Ecol Evol 22:621–624

Folke C, Hahn T, Olsson P, Norberg J (2005) Adaptive governance of social-ecological systems. Annu Rev Environ Resour 30:441–473

Foley JA, Ramankutty N, Brauman KA, Cassidy ES, Gerber JS, Johnston M, Mueller ND, O'Connell C, Ray DK, West PC (2011) Solutions for a cultivated planet. Nature 478:337–342

Franklin S (1995) Science as culture, cultures of science. Annu Rev Anthropol 24:163–184

Fuglie KO (2010) Total factor productivity in the global agricultural economy: evidence from FAO data. In: The shifting patterns of agricultural production and productivity worldwide. Midwest Agribusiness Trade Research and Information Center, Ames, pp 63–95

Funtowicz SO, Ravetz JR (1995) Science for the post normal age. In: Perspectives on ecological integrity. Springer, Dordrecht

Garnett T, Appleby MC, Balmford A, Bateman IJ, Benton TG, Bloomer P, Burlingame B, Dawkins M, Dolan L, Fraser D (2013) Sustainable intensification in agriculture: premises and policies. Science 341:33–34

Garnett T, Godfray C (2012) Sustainable intensification in agriculture. Navigating a course through competing food system priorities. In: Food climate research network and the Oxford Martin programme on the future of food, vol 51. University of Oxford, Oxford

Geels FW (2011) The multi-level perspective on sustainability transitions: responses to seven criticisms. Environ Innov Soc Trans 1:24–40

Gibson-Graham J (2014) Rethinking the economy with thick description and weak theory. Curr Anthropol 55:S147–S153

Gliessman SR (2014) Agroecology: the ecology of sustainable food systems. CRC press, Bosa Roca

Goddek S, Delaide B, Mankasingh U, Ragnarsdottir K, Jijakli H, Thorarinsdottir R (2015) Challenges of sustainable and commercial aquaponics. Sustainability 7:4199–4224

Godfray HCJ, Garnett T (2014) Food security and sustainable intensification. Philos Trans R Soc B 369:20120273

Grassini P, Eskridge KM, Cassman KG (2013) Distinguishing between yield advances and yield plateaus in historical crop production trends. Nat Commun 4:2918

Gunderson LH (2001) Panarchy: understanding transformations in human and natural systems. Island Press, Washington, DC

Gunderson R (2014) Problems with the defetishization thesis: ethical consumerism, alternative food systems, and commodity fetishism. Agric Hum Values 31:109–117

Gurian-Sherman D (2009) Failure to yield: evaluating the performance of genetically engineered crops, Union of Concerned Scientists. UCS Publications, Cambridge, MA

H2020-SCC-2016-2017 (smart and sustainable cities), Topic: SCC-02-2016-2017, Proposal number: 776528-1, Proposal acronym: proGIreg

Haenen I (2017) Small successes and big failures: lessons from the aquaponics facility at Uit Je Eigen Stad (UJES). 'Aquaponics.biz': A COST FA1305 Conference on Aquaponics SMEs, Rotterdam

Hall A, Bockett G, Taylor S, Sivamohan M, Clark N (2001) Why research partnerships really matter: innovation theory, institutional arrangements and implications for developing new technology for the poor. World Dev 29:783–797

Hamilton C (2013) Earthmasters: the dawn of the age of climate engineering. Yale University Press, New Haven

Hamilton C, Gemenne F, Bonneuil C (2015) The anthropocene and the global environmental crisis: rethinking modernity in a new epoch. Routledge, Oxon

Haraway D (2015) Anthropocene, capitalocene, plantationocene, chthulucene: making kin. Environ Humanit 6:159–165

Haraway DJ (1997) Modest-Witness@Second-Millennium.FemaleMan-Meets-OncoMouse : feminism and technoscience. Routledge, New York/London

Harding S (2004) Introduction: standpoint theory as a site of political, philosophic, and scientific debate. Routledge, New York

Harvey D (2007) A brief history of neoliberalism. Oxford University Press, New York

Healey P (2006) Urban complexity and spatial strategies: towards a relational planning for our times. Routledge, New York

Hinrichs CC (2003) The practice and politics of food system localization. J Rural Stud 19:33–45

Horkheimer M, Adorno TW (2002) Dialectic of enlightenment. Stanford University Press, Palo Alto

Hornborg A (2013) Technology as fetish: Marx, Latour, and the cultural foundations of capitalism. Theory Cult Soc 31:119–140

Hultman M (2013) The making of an environmental hero: a history of ecomodern masculinity, fuel cells and Arnold Schwarzenegger. Environ Humanit 2:79–99

Hunter MC, Smith RG, Schipanski ME, Atwood LW, Mortensen DA (2017) Agriculture in 2050: recalibrating targets for sustainable intensification. Bioscience 67:386–391

Jackson L, van Noordwijk M, Bengtsson J, Foster W, Lipper L, Pulleman M, Said M, Snaddon J, Vodouhe R (2010) Biodiversity and agricultural sustainagility: from assessment to adaptive management. Curr Opin Environ Sustain 2:80–87

Jahn T (2013) Sustainability science requires a critical orientation. GAIA 22:29–33

Jasanoff S (2007) Technologies of humility. Nature 450:33

Joly A, Junge R, Bardocz T (2015) Aquaponics business in Europe: some legal obstacles and solutions. Ecocycles 1:3–5

Junge R, König B, Villarroel M, Komives T, Jijakli MH (2017) Strategic points in aquaponics. Water 9:182

Kearney J (2010) Food consumption trends and drivers. Philos. Trans. R. Soc., B 365:2793–2807

Keil R (2007) Sustaining modernity, modernizing nature. In: The sustainable development paradox: urban political ecology in the US and Europe. Guilford Press, New York, pp 41–65

Kiers ET, Leakey RR, Izac A-M, Heinemann JA, Rosenthal E, Nathan D, Jiggins J (2008) Agriculture at a crossroads. Science 320:320–321

Kloppenburg J (1991) Social theory and the de/reconstruction of agricultural science: local knowledge for an alternative agriculture. Rural Sociol 56:519–548

Kläy A, Zimmermann AB, Schneider F (2015) Rethinking science for sustainable development: reflexive interaction for a paradigm transformation. Futures 65:72–85

Kőmíves T, Ranka J (2015) On the aquaponic corner section of our journal. Ecocycles 1:1–2

König B, Janker J, Reinhardt T, Villarroel M, Junge R (2018) Analysis of aquaponics as an emerging technological innovation system. J Clean Prod 180:232–243

Konig B, Junge R, Bittsanszky A, Villarroel M, Komives T (2016) On the sustainability of aquaponics. Ecocycles 2:26–32

Krueger R, Gibbs D (2007) The sustainable development paradox: urban political economy in the United States and Europe. Guilford Press, New York

Kuyper TW, Struik PC (2014) Epilogue: global food security, rhetoric, and the sustainable intensification debate. Curr Opin Environ Sustain 8:71–79

Laidlaw J, Magee L (2016) Towards urban food sovereignty: the trials and tribulations of community-based aquaponics enterprises in Milwaukee and Melbourne. Local Environ 21:573–590

Leach M, Rockström J, Raskin P, Scoones I, Stirling A, Smith A, Thompson J, Millstone E, Ely A, Arond E (2012) Transforming innovation for sustainability. Ecol Soc 17

Latour B (1986) Visualization and cognition. In: Knowledge and society, vol 6. Open University Press, Milton Keynes, pp 1–40

Latour B (1987) Science in action: how to follow scientists and engineers through society. Harvard University Press, Cambridge, MA

Latour B (1993) We have never been modern. Harvester Wheatsheaf, New York/London

Latour B (1996) Aramis, or, the love of technology. Harvard University Press, Cambridge, MA
Latour B (2004) Why has critique run out of steam? From matters of fact to matters of concern. Crit Inq 30:225–248
Latour B (2015) Telling friends from foes in the time of the Anthropocene. In: Hamilton C, Bonneuil C, Gemenne F (eds) The Anthropocene and the global environmental crisis: rethinking modernity in a new epoch. Routledge, Abingdon/Oxon
Latour B, Woolgar S (1986[1979]) Laboratory life: the construction of scientific facts. Princeton University Press, Princeton
Lave R, Mirowski P, Randalls S (2010) Introduction: STS and neoliberal science, vol 32. Sage, London, p 463
Law J, Williams RJ (1982) Putting facts together: a study of scientific persuasion. Soc Stud Sci 12:535–558
Lawrence RJ (2015) Advances in transdisciplinarity: epistemologies, methodologies and processes. Futures 65:1–9
Lawrence G, Richards C, Lyons K (2013) Food security in Australia in an era of neoliberalism, productivism and climate change. J Rural Stud 29:30–39
Leonard L (2013) Ecomodern discourse and localized narratives: waste policy, community mobilization and governmentality in Ireland. In: Organising waste in the City: international perspectives on narratives and practices. Policy Press, Chicago, pp 181–200
Lilley S, Papadopoulos D (2014) Material returns: cultures of valuation, biofinancialisation and the autonomy of politics. Sociology 48:972–988
Lobell DB, Schlenker W, Costa-Roberts J (2011) Climate trends and global crop production since 1980. Science 333:616–620
Loos J, Abson DJ, Chappell MJ, Hanspach J, Mikulcak F, Tichit M, Fischer J (2014) Putting meaning back into "sustainable intensification". Front Ecol Environ 12:356–361
Lorimer J (2017) The Anthropo-scene: a guide for the perplexed. Soc Stud Sci 47:117–142
Love DC, Fry JP, Li X, Hill ES, Genello L, Semmens K, Thompson RE (2015) Commercial aquaponics production and profitability: findings from an international survey. Aquaculture 435:67–74
Lundqvist J, de Fraiture C, Molden D (2008) Saving water: from field to fork: curbing losses and wastage in the food chain. Stockholm International Water Institute, Stockholm
LAWRENCE, R. J. 2015. Advances in transdisciplinarity: Epistemologies, methodologies and processes.Futures, 65, 1-9
Marcuse H (2013) One-dimensional man: studies in the ideology of advanced industrial society. Routledge, London
Martinez-Torres ME, Rosset PM (2010) La Vía Campesina: the birth and evolution of a transnational social movement. J Peasant Stud 37:149–175
Marzec RP (2014) Neoliberalism, Environmentality, and the specter of Sajinda khan. In: Di Leo JR, Mehan U (eds) Capital at the Brink: overcoming the destructive legacies of neoliberalism. Open Humanities Press, London
Mcmichael P (1994) The global restructuring of agro-food systems. Cornell University Press, Ithaca
Meadows DH, Meadows DL, Randers J, Behrens WW (1972) The limits to growth, vol 102. Universe Books, New York, p 27
Merchant C (1981) The death of nature: women, ecology, and scientific revolution. HarperOne, San Francisco
Miličić V, Thorarinsdottir R, Santos MD, Hančič MT (2017) Commercial aquaponics approaching the european market: to consumers' perceptions of aquaponics products in europe. Water 9:80
Miller TR, Neff MW (2013) De-facto science policy in the making: how scientists shape science policy and why it matters (or, why STS and STP scholars should socialize). Minerva 51:295–315
Miller TR, Wiek A, Sarewitz D, Robinson J, Olsson L, Kriebel D, Loorbach D (2014) The future of sustainability science: a solutions-oriented research agenda. Sustain Sci 9:239–246

Moore JW (2015) Capitalism in the web of life: ecology and the accumulation of capital. Verso Books, New York

Nally D (2011) The biopolitics of food provisioning. Trans Inst Br Geogr 36:37–53

Noordwijk V (2014) Climate change: agricultural mitigation. In: van Alfen NK (ed) Encyclopedia of agriculture and food systems. Elsevier, San Diego

Norton BG (2005) Sustainability: a philosophy of adaptive ecosystem management. University of Chicago Press, Chicago

Nugent RA (1999) Measuring the sustainability of urban agriculture. For hunger-proof cities. Sustainable urban food systems. IDRC, Ottawa, pp 95–99

Oldfield F, Richardson K, Schellnhuber H, Turner B II, Wasson R, Planet A (2004) Global change and the earth system. Springer, New York

Parfitt J, Barthel M, Macnaughton S (2010) Food waste within food supply chains: quantification and potential for change to 2050. Philosophical Transactions of the Royal Society B 365:3065–3081

Pearson CJ (2007) Regenerative, semiclosed systems: a priority for twenty-first-century agriculture. Bioscience 57:409

Pearson LJ, Pearson L, Pearson CJ (2010) Sustainable urban agriculture: stocktake and opportunities. Int J Agric Sustain 8:7–19

Pelletier N, Tyedmers P (2010) Forecasting potential global environmental costs of livestock production 2000–2050. Proc Natl Acad Sci 107:18371–18374

Perrot N, Trelea I-C, Baudrit C, Trystram G, Bourgine P (2011) Modelling and analysis of complex food systems: state of the art and new trends. Trends Food Sci Technol 22:304–314

Petersen B, Snapp S (2015) What is sustainable intensification? Views from experts. Land Use Policy 46:1–10

Pickering A (1992) Science as practice and culture. University of Chicago Press, Chicago

Postel SL (2003) Securing water for people, crops, and ecosystems: new mindset and new priorities. Nat Res Forum:89–98. Wiley Online Library

Potter C (2004) Multifunctionality as an agricultural and rural policy concept. Sustaining agriculture and the rural environment: governance, policy and multifunctionality. Edward Elgar, Northampton, pp 15–35

Power EM (1999) Combining social justice and sustainability for food security. In: For hunger-proof cities: sustainable urban food systems. International Development Research Centre, Ottawa, pp 30–37

Prado JR, Segers G, Voelker T, Carson D, Dobert R, Phillips J, Cook K, Cornejo C, Monken J, Grapes L (2014) Genetically engineered crops: from idea to product. Annu Rev Plant Biol 65:769

Rakocy JE, Masser MP, Losordo TM (2006) Recirculating aquaculture tank production systems: aquaponics—integrating fish and plant culture. SRAC Publication 454:1–16

Reynolds HL, Smith AA, Farmer JR (2014) Think globally, research locally: paradigms and place in agroecological research. Am J Bot 101:1631–1639

Robles AG (2015) Introducing ENoLL and its living lab community. European Network of Living Labs, Brussels

Rockstrom J, Steffen W, Noone K, Persson A, Chapin FS, Lambin E, Lenton TM, Scheffer M, Folke C, Schellnhuber HJ, Nykvist B, De Wit CA, Hughes T, van der Leeuw S, Rodhe H, Sorlin S, Snyder PK, Costanza R, Svedin U, Falkenmark M, Karlberg L, Corell RW, Fabry VJ, Hansen J, Walker B, Liverman D, Richardson K, Crutzen P, Foley J (2009) Planetary boundaries: exploring the safe operating space for humanity. Ecol Soc 14:32

Rockström J, Williams J, Daily G, Noble A, Matthews N, Gordon L, Wetterstrand H, Declerck F, Shah M, Steduto P, De Fraiture C, Hatibu N, Unver O, Bird J, Sibanda L, Smith J (2017) Sustainable intensification of agriculture for human prosperity and global sustainability. Ambio 46:4–17

Röling N (2009) Pathways for impact: scientists' different perspectives on agricultural innovation. Int J Agric Sustain 7:83–94

Rose N (2009) The politics of life itself: biomedicine, power, and subjectivity in the twenty-first century. Princeton University Press, Princeton

Savransky M (2013) An ecology of times: modern knowledge, non-modern temporalities. In: Movements in time: revolution, social justice and times of change, vol 265. Cambridge Scholars Publishing, Newcastle upon Tyne

Schmautz Z, Graber A, Jaenicke S, Goesmann A, Junge R, Smits TH (2017) Microbial diversity in different compartments of an aquaponics system. Arch Microbiol 199:613–620

Scott JC (2008) Weapons of the weak: everyday forms of peasant resistance. Yale University Press, New Haven

Schoenbach K (2001) Myths of media and audiences: inaugural lecture as professor of general communication science, University of Amsterdam. Eur J Commun 16:361–376

Sen A (2001) Development as freedom. Oxford Paperbacks, Oxford

Smil V (2011) Harvesting the biosphere: the human impact. Popul Dev Rev 37:613–636

Smith A, Stirling A (2007) Moving outside or inside? Objectification and reflexivity in the governance of socio-technical systems. J Environ Policy Plan 9:351–373

Soini K, Birkeland I (2014) Exploring the scientific discourse on cultural sustainability. Geoforum 51:213–223

Steffen W, Broadgate W, Deutsch L, Gaffney O, Ludwig C (2015a) The trajectory of the Anthropocene: the great acceleration. Anthropocene Rev 2:81–98

Steffen W, Richardson K, Rockström J, Cornell SE, Fetzer I, Bennett EM, Biggs R, Carpenter SR, De Vries W, De Wit CA (2015b) Planetary boundaries: guiding human development on a changing planet. Science 347:1259855

Steffen W, Sanderson RA, Tyson PD, Jäger J, Matson PA, Moore B III, Oldfield F, Richardson K, Schellnhuber HJ, Turner BL (2006) Global change and the earth system: a planet under pressure. Springer, New York

Stengers I (2015) In catastrophic times: resisting the coming barbarism. Open Humanities Press, London

Stengers I (2018) Another science is possible: a manifesto for slow science. Polity Press, Cambridge

Stern N (2008) The economics of climate change. Am Econ Rev 98:1–37

Stocker T (2014) Climate change 2013: the physical science basis: working group I contribution to the fifth assessment report of the intergovernmental panel on climate change. Cambridge University Press, Cambridge

Struik P (2006) Trends in agricultural science with special reference to research and development in the potato sector. Potato Res 49:5

Struik P, Kuyper T, Brussaard L, Leeuwis C (2014) Deconstructing and unpacking scientific controversies in intensification and sustainability: why the tensions in concepts and values? Curr Opin Environ Sustain 8:80–88

Struik PC, Kuyper TW (2014) Editorial overview: sustainable intensification to feed the world: concepts, technologies and trade-offs. Curr Opin Environ Sustain 8:vi–viii

Stuart T (2009) Waste: uncovering the global food scandal. WW Norton & Company, New York

Tegtmeier EM, Duffy MD (2004) External costs of agricultural production in the United States. Int J Agric Sustain 2:1–20

Thrift N (1999) Steps to an ecology of place. Human geography today. Polity Press, Cambridge, pp 295–322

Tilman D, Balzer C, Hill J, Befort BL (2011) Global food demand and the sustainable intensification of agriculture. Proc Natl Acad Sci 108:20260–20264

Tittonell P, Giller KE (2013) When yield gaps are poverty traps: the paradigm of ecological intensification in African smallholder agriculture. Field Crop Res 143:76–90

van der Ploeg JD, Ventura F (2014) Heterogeneity reconsidered. Curr Opin Environ Sustain 8:23–28

Vermeulen SJ, Campbell BM, Ingram JS (2012) Climate change and food systems. Annu Rev Environ Resour 37:195

Villarroel M, Junge R, Komives T, König B, Plaza I, Bittsánszky A, Joly A (2016) Survey of aquaponics in europe. Water 8:468

Wall D (2007) Realist utopias? Green alternatives to capitalism. Environ Politics 16:518–522

Weber KM, Rohracher H (2012) Legitimizing research, technology and innovation policies for transformative change: combining insights from innovation systems and multi-level perspective in a comprehensive 'failures' framework. Res Policy 41:1037–1047

Weis T (2010) The accelerating biophysical contradictions of industrial capitalist agriculture. J Agrar Chang 10:315–341

Wiek A, Ness B, Schweizer-Ries P, Brand FS, Farioli F (2012) From complex systems analysis to transformational change: a comparative appraisal of sustainability science projects. Sustain Sci 7:5–24

Williams JW, Jackson ST (2007) Novel climates, no-analog communities, and ecological surprises. Front Ecol Environ 5:475–482

Wilson GA (2001) From productivism to post-productivism… and back again? Exploring the (un) changed natural and mental landscapes of European agriculture. Trans Inst Br Geogr 26:77–102

Winner L (1986) The whale and the reactor: a search for limits in an age of high technology. University of Chicago Press, Chicago

Withers CW (2009) Place and the "Spatial Turn" in geography and in history. J Hist Ideas 70:637–658

Zalasiewicz J, Williams M, Steffen W, Crutzen P (2010) The new world of the Anthropocene. Environ Sci Technol 44:2228–2231

7

Plant Pathogens and Control Strategies in Aquaponics

Gilles Stouvenakers, Peter Dapprich, Sebastien Massart, and M. Haïssam Jijakli

Abstract Among the diversity of plant diseases occurring in aquaponics, soil-borne pathogens, such as *Fusarium* spp., *Phytophthora* spp. and *Pythium* spp., are the most problematic due to their preference for humid/aquatic environment conditions. *Phytophthora* spp. and *Pythium* spp. which belong to the Oomycetes pseudo-fungi require special attention because of their mobile form of dispersion, the so-called zoospores that can move freely and actively in liquid water. In coupled aquaponics, curative methods are still limited because of the possible toxicity of pesticides and chemical agents for fish and beneficial bacteria (e.g. nitrifying bacteria of the biofilter). Furthermore, the development of biocontrol agents for aquaponic use is still at its beginning. Consequently, ways to control the initial infection and the progression of a disease are mainly based on preventive actions and water physical treatments. However, suppressive action (suppression) could happen in aquaponic environment considering recent papers and the suppressive activity already highlighted in hydroponics. In addition, aquaponic water contains organic matter that could promote establishment and growth of heterotrophic bacteria in the system or even improve plant growth and viability directly. With regards to organic hydroponics (i.e. use of organic fertilisation and organic plant media), these bacteria could act as antagonist agents or as plant defence elicitors to protect plants from diseases. In the future, research on the disease suppressive ability of the aquaponic biotope must be increased, as well as isolation, characterisation and formulation of microbial plant pathogen antagonists. Finally, a good knowledge in the rapid identification of pathogens, combined with control methods and diseases monitoring, as recommended in integrated plant pest management, is the key to an efficient control of plant diseases in aquaponics.

G. Stouvenakers (✉) · S. Massart · M. H. Jijakli
Integrated and Urban Plant Pathology Laboratory, Université de Liège, Gembloux Agro-Bio Tech, Gembloux, Belgium
e-mail: g.stouvenakers@uliege.be; sebastien.massart@uliege.be; mh.jijakli@uliege.be

P. Dapprich
Department of Agriculture, Fachhochschule Südwestfalen University of Applied Sciences, Soest, Germany
e-mail: dapprich.peter@fh-swf.de

Keywords Pathogens · Fungi · Aquaponics · Biotope · Suppressive · Plant pest management · Biocontrol agents

Introduction

Nowadays, aquaponic systems are the core of numerous research efforts which aim at better understanding these systems and at responding to new challenges of food production sustainability (Goddek et al. 2015; Villarroel et al. 2016). The cumulated number of publications mentioning "aquaponics" or derived terms in the title went from 12 in early 2008 to 215 in 2018 (January 2018 Scopus database research results). In spite of this increasing number of papers and the large area of study topics they are covering, one critical point is still missing, namely plant pest management (Stouvenakers et al. 2017). According to a survey on EU Aquaponic Hub members, only 40% of practitioners have some notions about pests and plant pest control (Villarroel et al. 2016).

In aquaponics, the diseases might be similar to those found in hydroponic systems under greenhouse structures. Among the most problematic pathogens, in term of spread, are hydrophilic fungi or fungus-like protists which are responsible for root or collar diseases. To consider plant pathogen control in aquaponics, firstly, it is important to differentiate between coupled and decoupled systems. Decoupled systems allow disconnection between water from the fish and crop compartment (see Chapter). This separation allows the optimisation and a better control of different parameters (e.g. temperature, mineral or organic composition and pH) in each compartment (Goddek et al. 2016; Monsees et al. 2017). Furthermore, if the water from the crop unit does not come back to the fish part, the application of phytosanitary treatments (e.g. pesticides, biopesticides and chemical disinfection agents) could be allowed here. Coupled systems are built in one loop where water recirculates in all parts of the system. However, in coupled systems, plant pest control is more difficult due to the both presence of fish and beneficial microorganisms which transform fish sludge into plant nutrients. Their existence limits or excludes the application of already available disinfecting agents and chemical treatments. Furthermore no pesticides or biopesticides have been specifically developed for aquaponics (Rakocy et al. 2006; Rakocy 2012; Somerville et al. 2014; Bittsanszky et al. 2015; Nemethy et al. 2016; Sirakov et al. 2016). Control measures are consequently mainly based on non-curative physical practices (Nemethy et al. 2016; Stouvenakers et al. 2017).

On the other hand, recent studies highlight that aquaponic plant production offers similar yields when compared to hydroponics although concentrations of mineral plant nutrients are lower in aquaponic water. Furthermore, when aquaponic water is complemented with some minerals to reach hydroponic concentrations of mineral nutritive elements, even better yields can be observed (Pantanella et al. 2010; Pantanella et al. 2015; Delaide et al. 2016; Saha et al. 2016; Anderson et al. 2017;

Wielgosz et al. 2017; Goddek and Vermeulen 2018). Moreover, some informal observations from practitioners in aquaponics and two recent scientific studies (Gravel et al. 2015; Sirakov et al. 2016) report the possible presence of beneficial compounds and/or microorganisms in the water that could play a role in biostimulation and/or have antagonistic (i.e. inhibitory) activity against plant pathogens. Biostimulation is defined as the enhancement of plant quality traits and plant tolerance against abiotic stress using any microorganism or substance.

With regard to these aspects, this chapter has two main objectives. The first is to give a review of microorganisms involved in aquaponic systems with a special focus on plant pathogenic and plant beneficial microorganisms. Factors influencing these microorganisms will be also considered (e.g. organic matter). The second is to review available methods and future possibilities in plant diseases control.

Microorganisms in Aquaponics

Microorganisms are present in the entire aquaponics system and play a key role in the system. They are consequently found in the fish, the filtration (mechanical and biological) and the crop parts. Commonly, the characterisation of microbiota (i.e. microorganisms of a particular environment) is carried out on circulating water, periphyton, plants (rhizosphere, phyllosphere and fruit surface), biofilter, fish feed, fish gut and fish faeces. Up until now, in aquaponics, most of microbial research has focused on nitrifying bacteria (Schmautz et al. 2017). Thus, the trend at present is to characterise microorganisms in all compartments of the system using modern sequencing technologies. Schmautz et al. (2017) identified the microbial composition in different parts of the system, whereas Munguia-Fragozo et al. (2015) give perspectives on how to characterize aquaponics microbiota from a taxonomical and functional point of view by using cutting-edge technologies. In the following sub-sections, focus will be only brought on microorganisms interacting with plants in aquaponic systems organised into plant beneficial and plant pathogenic microorganisms.

Plant Pathogens

Plant pathogens occurring in aquaponic systems are theoretically those commonly found in soilless systems. A specificity of aquaponic and hydroponic plant culture is the continuous presence of water in the system. This humid/aquatic environment suits almost every plant pathogenic fungus or bacteria. For root pathogens some are particularly well adapted to these conditions like pseudo-fungi belonging to the taxa of Oomycetes (e.g. root rot diseases caused by *Pythium* spp. and *Phytophthora* spp.) which are able to produce a motile form of dissemination called zoospores. These zoospores are able to move actively in liquid water and thus are able to spread over the entire system extremely quickly. Once a plant is infected, the disease can rapidly spread out the system, especially because of the water's recirculation (Jarvis 1992; Hong and Moorman 2005; Sutton et al. 2006; Postma et al. 2008; Vallance et al. 2010; Rakocy 2012; Rosberg 2014; Somerville et al. 2014). Though Oomycetes are among the most prevalent pathogens detected during root diseases, they often form a complex with other pathogens. Some *Fusarium* species (with existence of species well adapted to aquatic environment) or species from the genera *Colletotrichum*, *Rhizoctonia* and *Thielaviopsis* can be found as part of these complexes and can also cause significant damage on their own (Paulitz and Bélanger 2001; Hong and Moorman 2005; Postma et al. 2008; Vallance et al. 2010). Other fungal genera like *Verticillium* and *Didymella*, but also bacteria, such as *Ralstonia*, *Xanthomonas*, *Clavibacter*, *Erwinia* and *Pseudomonas*, as well as viruses (e.g. tomato mosaic, cucumber mosaic, melon necrotic spot virus, lettuce infectious virus and tobacco necrosis), can be detected in hydroponics or irrigation water and cause vessel, stem, leaf or fruit damage (Jarvis 1992; Hong and Moorman 2005). However note that not all microorganisms detected are damaging or lead to symptoms in the crop. Even species of the same genus can be either harmful or beneficial (e.g. *Fusarium*, *Phoma*, *Pseudomonas*). Disease agents discussed above are mainly pathogens linked to water recirculation but can be identified in greenhouses also. Section shows the results of the first international survey on plant diseases occurring specifically in aquaponics, while Jarvis (1992) and Albajes et al. (2002) give a broader view of occurring pathogens in greenhouse structures.

In hydroponics or in aquaponic systems, plants generally grow under greenhouse conditions optimized for plant production, especially for large-scale production where all the environmental parameters are computer managed (Albajes et al. 2002; Vallance et al. 2010; Somerville et al. 2014; Parvatha Reddy 2016). However, optimal conditions for plant production can also be exploited by plant pathogens. In fact, these structures generate warm, humid, windless and rain-free conditions that can encourage plant diseases if they are not correctly managed (ibid.). To counteract this, compromises must be made between optimal plant conditions and disease prevention (ibid.). In the microclimate of the greenhouse, an inappropriate management of the vapour-pressure deficit can lead to the formation of a film or a drop of water on the plants surface. This often promotes plant pathogen development. Moreover, to maximise the yield in commercial hydroponics, some other parameters (e.g. high plant density, high fertilisation, to extend the period production) can enhance the susceptibility of plants to develop diseases (ibid.).

The question now is to know by which route the initial inoculum (i.e. the first step in an epidemiological cycle) is brought into the system. The different steps in plant disease epidemiological cycle (EpC) are represented in Fig. In aquaponics, as in greenhouse hydroponic culture, it can be considered that entry of pathogens could be linked to water supply, introduction of infected plants or seeds, the growth material (e.g. reuse of the media), air exchange (dust and particles carriage), insects (vectors of diseases and particles carriage) and staff (tools and clothing) (Paulitz and Bélanger 2001; Albajes et al. 2002; Hong and Moorman 2005; Sutton et al. 2006; Parvatha Reddy 2016).

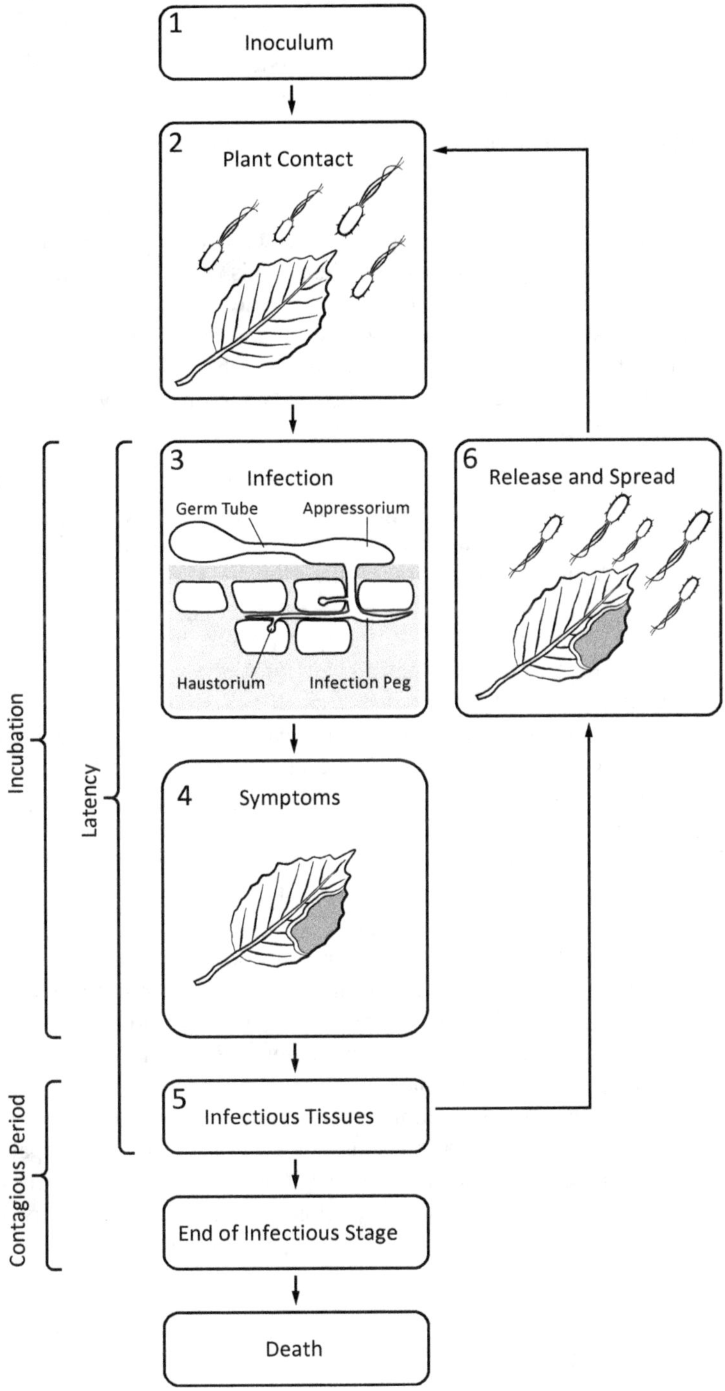

Fig. Basic steps (*1* to *6*) in plant disease epidemiological cycle (EpC) according to Lepoivre (2003). (*1*) Arrival of the pathogen inoculum, (*2*) contact with the host plant, (*3*) tissues penetration and infection process by the pathogen, (*4*) symptoms development, (*5*) plant tissues that become infectious, (*6*) release and spread of infectious form of dispersion

Once the inoculum is in contact with the plant (step 2 in the EpC), several cases of infection (step 3 in the EpC) are possible (Lepoivre 2003):

- The pathogen-plant relationship is incompatible (non-host relation) and disease does not develop.
- There is a host relation but the plant does not show symptoms (the plant is tolerant).
- The pathogen and the plant are compatible but defence response is strong enough to inhibit the progression of the disease (the plant is resistant: interaction between host resistance gene and pathogen avirulence gene).
- The plant is sensitive (host relation without gene for gene recognition), and the pathogen infects the plant, but symptoms are not highly severe (step 4 in the EpC).
- And lastly, the plant is sensitive and disease symptoms are visible and severe (step 4 in the EpC).

Regardless of the degree of resistance, some environmental conditions or factors can influence the susceptibility of a plant to be infected, either by a weakening of the plant or by promoting the growth of the plant pathogen (Colhoun 1973; Jarvis 1992; Cherif et al. 1997; Alhussaen 2006; Somerville et al. 2014). The main environmental factors influencing plant pathogens and disease development are temperature, relative humidity (RH) and light (ibid.). In hydroponics, temperature and oxygen concentrations within the nutrient solution can constitute additional factors (Cherif et al. 1997; Alhussaen 2006; Somerville et al. 2014). Each pathogen has its own preference of environmental conditions which can vary during its epidemiologic cycle. But in a general way, high humidity and temperature are favourable to the accomplishment of key steps in the pathogen's epidemic cycle such as spore production or spore germination (Fig., step 5 in the EpC) (Colhoun 1973; Jarvis 1992; Cherif et al. 1997; Alhussaen 2006; Somerville et al. 2014). Colhoun (1973) sums up the effects of the various factors promoting plant diseases in soil, whereas Table shows the more specific or adding factors that may encourage plant pathogen development linked to aquaponic greenhouse conditions.

In the epidemiological cycle, once the infective stage is reached (step 5 in the EpC), the pathogens can spread in several ways (Fig, step 6 in the EpC) and infect other plants. As explained before, root pathogens belonging to Oomycetes taxa can actively spread in the recirculating water by zoospores release (Alhussaen 2006; Sutton et al. 2006). For other fungi, bacteria and viruses responsible for root or aerial diseases, the dispersion of the causal agent can occur by propagation of infected material, mechanical wounds, infected tools, vectors (e.g. insects) and particles (e.g. spores and propagules) ejection or carriage allowed by drought, draughts or water splashes (Albajes et al. 2002; Lepoivre 2003).

Survey on Aquaponic Plant Diseases

During January 2018, the first international survey on plant diseases was made among aquaponics practitioner members of the COST FA1305, the American Aquaponics Association and the EU Aquaponics Hub. Twenty-eight answers were

Table Adding factors encouraging plant pathogen development under aquaponic greenhouse structure compared to classical greenhouse culture

Promoting factor	Profiting to	Causes	References
Nutrient film technique (NFT), Deep flow technique (DFT)	*Pythium* spp., *Fusarium* spp.	Easy spread by water recirculation; possibility of post contamination after a disinfection step; poor content in oxygen in the nutrient solution	Koohakan et al. (2004) and Vallance et al. (2010)
Inorganic media (e.g. rockwool)	Higher content in bacteria (no information about their possible pathogenicity)	Unavailable organic compounds in the media	Khalil and Alsanius (2001), Koohakan et al. (2004), Vallance et al. (2010)
Organic media (e.g. coconut fibre and peat)	Higher content in fungi; higher content in *Fusarium* spp. for coconut fibre	Available organic compounds in the media	Koohakan et al. (2004), Khalil et al. (2009), and Vallance et al. (2010)
Media with high water content and low content in oxygen (e.g. rockwool)	*Pythium* spp.	Zoospores mobility; plant stress	Van Der Gaag and Wever (2005), Vallance et al. (2010), and Khalil and Alsanius (2011)
Media allowing little water movement (e.g. rockwool)	*Pythium* spp.	Better condition for zoospores dispersal and chemotaxis movement; no loss of zoospore flagella	Sutton et al. (2006)
High temperature and low concentration of DO in the nutrient solution	*Pythium* spp.	Plant stressed and optimal condition for *Pythium* growth	Cherif et al. (1997), Sutton et al. (2006), Vallance et al. (2010), and Rosberg (2014)
High host plant density and resulting microclimate	Pathogens growth; diseases spread	Warm and humid environment	Albajes et al. (2002) and Somerville et al. (2014)
Deficiencies, excess or imbalance of macro/micronutrients	Fungi, viruses and bacteria	Plant physiological modifications (e.g. action on defence response, transpiration, integrity of cell walls); plant morphological modifications (e.g. higher susceptibility to pathogens, attraction of pests); nutrient resources in host tissues for pathogens; direct action on the pathogen development cycle	Colhoun (1973), Snoeijers and Alejandro (2000), Mitchell et al. (2003), Dordas (2008), Veresoglou et al. (2013), Somerville et al. (2014), and Geary et al. (2015)

received describing 32 aquaponic systems from around the world (EU, 21; North America, 5; South America, 1; Africa, 4; Asia, 1). The first finding was the small response rate. Among the possible explanations for the reluctance to reply to the questionnaire was that practitioners did not feel able to communicate about plant pathogens because of a lack of knowledge on this topic. This had already been observed in the surveys of Love et al. (2015) and Villarroel et al. (2016).

Key information obtained from the survey are:

- 84.4% of practitioners observe disease in their system.
- 78.1% cannot identify the causal agent of a disease.
- 34.4% do not apply disease control measures.
- 34.4% use physical or chemical water treatment.
- 6.2% use pesticides or biopesticides in coupled aquaponic system against plant pathogens.

These results support the previous arguments saying that aquaponic plants do get diseases. Yet, practitioners suffer from a lack of knowledge about plant pathogens and disease control measures actually used are essentially based on non-curative actions (90.5% of cases).

In the survey, a listing of plant pathogens occurring in their aquaponic system was provided. Table shows the results of this identification. To remedy the lack of practitioner's expertise about plant disease diagnostics, a second survey version was

Table Results of the first identifications of plant pathogens in aquaponics from the 2018 international survey analysis and from existing literature

Plant host	Plant pathogen	References or survey results
Allium schoenoprasu	*Pythium* sp.[(b)]	Survey
Beta vulgaris (swiss chard)	*Erysiphe betae*[(a)]	Survey
Cucumis sativus	*Podosphaera xanthii*[(a)]	Survey
Fragaria spp.	*Botrytis cinerea*[(a)]	Survey
Lactuca sativa	*Botrytis cinerea*[(a)]	Survey
	Bremia lactucae[(a)]	Survey
	Fusarium sp.[(b)]	Survey
	Pythium dissotocum[(b)]	Rakocy (2012)
	Pythium myriotylum[(b)]	Rakocy (2012)
	Sclerotinia sp.[(a)]	Survey
Mentha spp.	*Pythium* sp.[(b)]	Survey
Nasturtium officinale	*Aspergillus sp.*[(a)]	Survey
Ocimum basilicum	*Alternaria* sp.[(a)]	Survey
	Botrytis cinerea[(a)]	Survey
	Pythium sp.[(b)]	Survey
	Sclerotinia sp.[(a)]	Survey
Pisum sativum	*Erysiphe pisi*[(a)]	Survey
Solanum lycopersicum	*Pseudomononas solanacearum*[(a)]	McMurty et al. (1990)
	Phytophthora infestans[(a)]	Survey

Plant pathogens identified by symptoms in the aerial plant part are annotated by (a) and in root part by (b) in exponent

sent with the aim to identify symptoms without disease name linkage (Table). Table mainly identifies diseases with specific symptoms, i.e. symptoms that can be directly linked to a plant pathogen. It is the case of *Botrytis cinerea* and its typical grey mould, powdery mildew (*Erysiphe* and *Podosphaera* genera in the table) and its white powdery mycelium/conidia, and lastly *Sclerotinia* spp. and its sclerotia production. The presence of 3 plant pathologists in the survey respondents expands the list, with the identification of some root pathogens (e.g. *Pythium* spp.). General symptoms that are not specific enough to be directly related to a pathogen without further verification (see diagnosis in Section) are consequently found in Table. But it is important to highlight that most of the symptoms observed in this table could also be the consequence of abiotic stresses. Foliar chlorosis is one of the most explicit examples because it can be related to a large number of pathogens (e.g. for lettuces: *Pythium* spp., *Bremia lactucae*, *Sclerotinia* spp., beet western yellows virus), to environmental conditions (e.g. temperature excess) and to mineral deficiencies (nitrogen, magnesium, potassium, calcium, sulfur, iron, copper, boron, zinc, molybdenum) (Lepoivre 2003; Resh 2013).

Table Review of occurring symptoms in aquaponics from the 2018 international survey analysis

Symptoms	Plants species
Foliar chlorosis	*Allium schoenoprasum* ^{1}, *Amaranthus viridis* ^{1}, *Coriandrum sativum* ^{1}, *Cucumis sativus* ^{1}, Ocimum *basilicum* ^{6}, *Lactuca sativa* ^{4}, *Mentha* spp. ^{2}, *Petroselinum crispum* ^{1}, *Spinacia oleracea* ^{2}, *Solanum lycopersicum* ^{1}, *Fragaria* spp. ^{1}
Foliar necrosis	*Mentha* spp. ^{2}, *Ocimum basilicum* ^{1},
Stem necrosis	*Solanum lycopersicum* ^{1},
Collar necrosis	*Ocimum basilicum* ^{1}
Foliar Mosaic	*Cucumis sativus* ^{1}, *Mentha* spp. ^{1}, *Ocimum basilicum* ^{1},
Foliar wilting	*Brassica oleracea* Acephala group ^{1}, *Lactuca sativa* ^{1}, *Mentha* spp. ^{1}, *Cucumis sativus* ^{1}, *Ocimum basilicum* ^{1}, *Solanum lycopersicum* ^{1}
Foliar, stem and collar mould	*Allium schoenoprasum* ^{1}, *Capsicum annuum* ^{1}, *Cucumis sativus* ^{1}, *Lactuca sativa* ^{2}, *Mentha* spp. ^{2}, *Ocimum basilicum* ^{4}, *Solanum lycopersicum* ^{1}
Foliar spots	*Capsicum annuum* ^{1}, *Cucumis sativus* ^{1}, *Lactuca sativa* ^{2}, *Mentha* spp. ^{1}, *Ocimum basilicum* ^{5}
Damping off	*Spinacia oleracea* ^{1}, *Ocimum basilicum* ^{1}, *Solanum lycopersicum* ^{1}, seedlings in general ^{5}
Crinkle	*Beta vulgaris* (swiss chard) ^{1}, *Capsicum annuum* ^{1}, *Lactuca sativa* ^{1}, *Ocimum basilicum* ^{1}
Browning or decaying root	*Allium schoenoprasum* ^{1}, *Amaranthus viridis* ^{1}, *Beta vulgaris* (swiss chard) ^{1}, *Coriandrum sativum* ^{1}, *Lactuca sativa* ^{1}, *Mentha* spp. ^{2}, *Ocimum basilicum* ^{2}, *Petroselinum crispum* ^{2}, *Solanum lycopersicum* ^{1}, *Spinacia oleracea* ^{1}

Numbers in exponent represent the occurrence of the symptom for a specific plant on a total of 32 aquaponic systems reviewed

Beneficial Microorganisms in Aquaponics: The Possibilities

As explained in the introduction, several publications focused on bacteria involved in the nitrogen cycle, while others already emphasise the potential presence of beneficial microorganisms interacting with plant pathogens and/or plants (Rakocy 2012; Gravel et al. 2015; Sirakov et al. 2016). This section reviews the potential of plant beneficial microorganisms involved in aquaponics and their modes of action.

Sirakov et al. (2016) screened antagonistic bacteria against *Pythium ultimum* isolated from an aquaponic system. Among the 964 tested isolates, 86 showed a strong inhibitory effect on *Pythium ultimum* in vitro. Further research must be achieved to taxonomically identify these bacteria and evaluate their potential in in vivo conditions. The authors assume that many of these isolates belong to the genus *Pseudomonas*. Schmautz et al. (2017) came to the same conclusion by identifying *Pseudomonas* spp. in the rhizosphere of lettuce. Antagonistic species of the genus *Pseudomonas* were able to control plant pathogens in natural environments (e.g. in suppressive soils) while this action is also affected by environmental conditions. They can protect plants against pathogens either in an active or a passive way by eliciting a plant defence response, playing a role in plant growth promotion, compete with pathogens for space and nutrients (e.g. iron competition by release of iron-chelating siderophores), and/or finally by production of antibiotics or antifungal metabolites such as biosurfactants (Arras and Arru 1997; Ganeshan and Kumar 2005; Haas and Défago 2005; Beneduzi et al. 2012; Narayanasamy 2013). Although no identification of microorganisms was done by Gravel et al. (2015)), they report that fish effluents have the capacity to stimulate plant growth, decrease the mycelial growth of *Pythium ultimum* and *Pythium oxysporum* in vitro and reduce the colonization of tomato root by these fungi.

Information about the possible natural plant protection capacity of aquaponic microbiota is scarce, but the potential of this protective action can be envisaged with regard to different elements already known in hydroponics or in recirculated aquaculture. A first study was conducted in 1995 on suppressive action or suppressiveness promoted by microorganisms in soilless culture (McPherson et al. 1995). Suppressiveness in hydroponics, here defined by Postma et al. (2008)), has been "referred to the cases where (i) the pathogen does not establish or persist; or (ii) establishes but causes little or no damage". The suppressive action of a milieu can be related to the abiotic environment (e.g. pH and organic matter). However, in most situations, it is considered to be related directly or indirectly to microorganisms' activity or their metabolites (James and Becker 2007). In soilless culture, the suppressive capacity shown by water solution or the soilless media is reviewed by Postma et al. (2008) and Vallance et al. (2010). In these reviews, microorganisms responsible for this suppressive action are not clearly identified. In contrast, plant pathogens like *Phytophthora cryptogea*, *Pythium* spp., *Pythium aphanidermatum* and *Fusarium oxysporum* f.sp. *radicis-lycopersici* controlled or suppressed by the natural microbiota are exhaustively described. In the various articles reviewed by

Postma et al. (2008) and Vallance et al. (2010), microbial involvement in the suppressive effect is generally verified via a destruction of the microbiota of the soilless substrate by sterilisation first and eventually followed by a re-inoculation. When compared with an open system without recirculation, suppressive activity in soilless systems could be explained by the water recirculation (McPherson et al. 1995; Tu et al. 1999, cited by Postma et al. 2008) which could allow a better development and spread of beneficial microorganisms (Vallance et al. 2010).

Since 2010, suppressiveness of hydroponic systems has been generally accepted and research topics have been more driven on isolation and characterization of antagonistic strains in soilless culture with *Pseudomonas* species as main organisms studied. If it was demonstrated that soilless culture systems can offer suppressive capacity, there is no similar demonstration of such activity in aquaponics systems. However, there is no empiric indication that it should not be the case. This optimism arises from the discoveries of Gravel et al. (2015) and Sirakov et al. (2016) described in the second paragraph of this section. Moreover, it has been shown in hydroponics (Haarhoff and Cleasby 1991 cited by Calvo-bado et al. 2003; Van Os et al. 1999) but also in water treatment for human consumption (reviewed by Verma et al. 2017) that slow fi ltration (described in Section) and more precisely slow sand fi ltration can also act against plant pathogens by a microbial suppressive action in addition to other physical factors. In hydroponics, slow filtration has been demonstrated to be effective against the plant pathogens reviewed in Table. It is assumed that the microbial suppressive activity in the filters is most probably due to species of *Bacillus* and/or *Pseudomonas* (Brand 2001; Déniel et al. 2004; Renault et al. 2007; Renault et al. 2012). The results of Déniel et al. (2004) suggest that in hydroponics, the mode of action of *Pseudomonas* and *Bacillus* relies on competition for nutrients and antibiosis, respectively. However, additional modes of action could be present for these two genera as already explained for *Pseudomonas* spp. *Bacillus* species can, depending on the environment, act either indirectly by plant biostimulation or elicitation of plant defences or directly by antagonism via production of antifungal and/or antibacterial substances. Cell wall-degrading enzymes, bacteriocins, and antibiotics, lipopeptides (i.e. biosurfactants), are identified as key molecules for the latter action (Pérez-García et al. 2011; Beneduzi et al. 2012;

Table Review of plant pathogens effectively removed by slow filtration in hydroponics

Plant pathogens	References
Xanthomonas campestris pv. *Pelargonii*	Brand (2001)
Fusarium oxysporum	Wohanka (1995), Ehret et al. (1999) cited by Ehret et al. (2001), van Os et al. (2001), Déniel et al. (2004), and Furtner et al. (2007)
Pythium spp.	Déniel et al. (2004)
Pythium aphanidermatum	Ehret et al. (1999) cited by Ehret et al. (2001), and Furtner et al. (2007)
Phytophthora cinnamomi	Van Os et al. (1999), 4 references cited by Ehret et al. (2001)
Phytophthora cryptogea	Calvo-bado et al. (2003)
Phytophthora cactorum	Evenhuis et al. (2014)

Narayanasamy 2013). All things considered, the functioning of a slow filter is not so different from the functioning of some biofilters used in aquaponics. Furthermore, some heterotrophic bacteria like *Pseudomonas* spp. were already identified in aquaponics biofilters (Schmautz et al. 2017). This is in accordance with the results of other researchers who frequently detected *Bacillus* and/or *Pseudomonas* in RAS (recirculated aquaculture system) biofilters (Tal et al. 2003; Sugita et al. 2005; Schreier et al. 2010; Munguia-Fragozo et al. 2015; Rurangwa and Verdegem 2015). Nevertheless, up until now, no study about the possible suppressiveness in aquaponic biofilters has been carried out.

Protecting Plants from Pathogens in Aquaponics

At the moment aquaponic practitioners operating a coupled system are relatively helpless against plant diseases when they occur, especially in the case of root pathogens. No pesticide nor biopesticide is specifically developed for aquaponic use (Rakocy 2007; Rakocy 2012; Somerville et al. 2014; Bittsanszky et al. 2015; Sirakov et al. 2016). In brief, curative methods are still lacking. Only Somerville et al. (2014) list the inorganic compounds that may be used against fungi in aquaponics. In any case, an appropriate diagnostic of the pathogen(s) causing the disease is mandatory in order to identify the target(s) for curative measures. This diagnosis requires good expertise in terms of observation capacity, plant pathogen cycle understanding and analysis of the situation. However, in case of generalist (not specific) symptoms and depending on the degree of accuracy needed, it is often necessary to use laboratory techniques to validate the hypothesis with respect to the causal agent (Lepoivre 2003). Postma et al. (2008) reviewed the different methods to detect plant pathogens in hydroponics, and four groups were identified:

1. Direct macroscopic and microscopic observation of the pathogen
2. Isolation of the pathogen
3. Use of serological methods
4. Use of molecular methods

Non-biological Methods of Protection

Good agricultural practices (GAP) for plant pathogens control are the various actions aiming to limit crop diseases for both yield and quality of produce (FAO 2008). GAP transposable to aquaponics are essentially non-curative physical or cultivation practices that can be divided in preventive measures and water treatment.

Preventive Measures

Preventive measures have two distinct purposes. The first is to avoid the entry of the pathogen inoculum into the system and the second is to limit (i) plant infection,

(ii) development and (iii) spread of the pathogen during the growing period. Preventive measures aiming to avoid the entry of the initial inoculum in the greenhouse are, for example, a fallow period, a specific room for sanitation, room sanitation (e.g. plant debris removal and surface disinfection), specific clothes, certified seeds, a specific room for plant germination and physical barriers (against insect vectors) (Stanghellini and Rasmussen 1994; Jarvis 1992; Albajes et al. 2002; Somerville et al. 2014; Parvatha Reddy 2016). Among the most important practices used for the second type of preventive measures are, the use of resistant plant varieties, tools disinfection, avoidance of plant abiotic stresses, good plant spacing, avoidance of algae development and environmental conditions management. The last measure, i.e. environmental conditions management, means to control all greenhouse parameters in order to avoid or limit diseases by intervening in their biological cycle (ibid.). Generally, in large-scale greenhouse structures, computer software and algorithms are used to calculate the optimal parameters allowing both plant production and disease control. The parameters measured, among others, are temperature (of the air and the nutrient solution), humidity, vapour pressure deficit, wind speed, dew probability, leaf wetness and ventilation (ibid.). The practitioner acts on these parameters by manipulating the heating, the ventilation, the shading, the supplement of lights, the cooling and the fogging (ibid.).

Water Treatment

Physical water treatments can be employed to control potential water pathogens. Filtration (pore size less than 10 μm), heat and UV treatments are among the most effective to eliminate pathogens without harmful effects on fish and plant health (Ehret et al. 2001; Hong and Moorman 2005; Postma et al. 2008; Van Os 2009; Timmons and Ebeling 2010). These techniques allow the control of disease outbreaks by decreasing the inoculum, the quantity of pathogens and their proliferation stages in the irrigation system (ibid.). Physical disinfection decreases water pathogens to a certain level depending on the aggressiveness of the treatment. Generally, the target of heat and UV disinfection is the reduction of the initial microorganisms population by 90–99.9% (ibid.). The filtration technique most used is slow filtration because of its reliability and its low cost. The substrates of filtration generally used are sand, rockwool or pozzolana (ibid.). Filtration efficiency is essentially dependent on pore size and flow. To be effective as disinfection treatment, the filtration needs to be achieved with a pore size less than 10 μm and a flow rate of 100 $l/m^2/h$, even if less binding parameters show satisfactory performances (ibid.). Slow filtration does not eliminate all of the pathogens; more than 90% of the total aerobic bacteria remain in the effluent (ibid.). Nevertheless, it allows a suppression of plant debris, algae, small particles and some soil-borne diseases such as *Pythium* and *Phytophthora* (the efficiency is genus dependent). Slow filters do not act only by physical action but also show a microbial suppressive activity, thanks to antagonistic microorganisms, as discussed in Section (Hong and Moorman 2005 ; Postma et al. 2008 ; Van Os 2009; Vallance et al. 2010). Heat treatment is very effective against plant pathogens. However it requires temperatures reaching 95 ° C during at least 10 seconds to suppress all kind of pathogens, viruses included. This practice consumes a lot of energy and imposes water cooling (heat exchanger and transitional tank) before reinjection of the treated water back into the irrigation loop. In addition, it has the disadvantage of killing all microorganisms including the beneficial ones (Hong and Moorman 2005; Postma et al. 2008; Van Os 2009). The last technique and probably the most applied is UV disinfection. 20.8% of EU Aquaponics Hub practitioners use it (Villarroel et al. 2016). UV radiation has a wavelength of 200 to 280 nm. It has a detrimental effect on microorganisms by direct damage of the DNA. Depending on the pathogen and the water turbulence, the energy dose varies between 100 and 250 mJ/cm^2 to be effective (Postma et al. 2008; Van Os 2009).

Physical water treatments eliminate the most of the pathogens from the incoming water but they cannot eradicate the disease when it is already present in the system. Physical water treatment does not cover all the water (especially the standing water zone near the roots), nor the infected plant tissue. For example, UV treatments often fail to suppress *Pythium* root rot (Sutton et al. 2006). However, if physical water treatment allows a reduction of plant pathogens, theoretically, they also have an effect on nonpathogenic microorganisms potentially acting on disease suppression. In reality, heat and UV treatments create a microbiological vacuum, whereas slow filtration produces a shift in effluent microbiota composition resulting in a higher disease suppression capacity (Postma et al. 2008; Vallance et al. 2010). Despite the fact that UV and heat treatment in hydroponics eliminate more than 90% of microorganisms in the recirculating water, no diminution of the disease suppressiveness was observed. This was probably due to a too low quantity of water treated and a re-contamination of the water after contact with the irrigation system, roots and plant media (ibid.).

Aquaponic water treatment by means of chemicals is limited in continuous application. Ozonation is a technique used in recirculated aquaculture and in hydroponics. Ozone treatment has the advantage to eliminate all pathogens including viruses in certain conditions and to be rapidly decomposed to oxygen (Hong and Moorman 2005; Van Os 2009; Timmons and Ebeling 2010; Gonçalves and Gagnon 2011). However it has several disadvantages. Introducing ozone in raw water can produce by-products oxidants and significant amount of residual oxidants (e.g. brominated compound and haloxy anions that are toxic for fish) that need to be removed, by UV radiation, for example, prior to return to the fish part (reviewed by Gonçalves and Gagnon 2011). Furthermore, ozone treatment is expensive, is irritant for mucous membranes in case of human exposure, needs contact periods of 1 to 30 minutes at a concentration range of 0.1–2.0 mg/L, needs a temporal sump to reduce completely from O_3 to O_2 and can oxidize elements present in the nutrient solution, such as iron chelates, and thus makes them unavailable for plants (Hong and Moorman 2005; Van Os 2009; Timmons and Ebeling 2010; Gonçalves and Gagnon 2011).

Biological Methods of Protection

In hydroponics, numerous scientific papers review the use of antagonistic microorganisms (i.e. able to inhibit other organisms) to control plant pathogens but until now no research has been carried out for their use in aquaponics. The mode of action of these antagonistic microorganisms is according to Campbell (1989)), Whipps (2001) and Narayanasamy (2013) grouped in:

1. Competition for nutrients and niches
2. Parasitism
3. Antibiosis
4. Induction of diseases resistance in plants

The experiments introducing microorganisms in aquaponic systems have been focused on the increase of nitrification by addition of nitrifying bacteria (Zou et al. 2016) or the use of plant growth promoters (PGPR) such as *Azospirillum brasilense* and *Bacillus* spp. to increase plant performance (Mangmang et al. 2014; Mangmang et al. 2015a; Mangmang et al. 2015b; Mangmang et al. 2015c; da Silva Cerozi and Fitzsimmons 2016; Bartelme et al. 2018). There is now an urgent need to work on biocontrol agents (BCA) against plant pathogens in aquaponics with regard to the restricted use of synthetic curative treatments, the high value of the culture and the increase of aquaponic systems in the world. BCA are defined, in this context, as viruses, bacteria and fungi exerting antagonistic effects on plant pathogens (Campbell 1989; Narayanasamy 2013).

Generally, the introduction of a BCA is considered to be easier in soilless systems. In fact, the hydroponic root environment is more accessible than in soil and the microbiota of the substrate is also unbalanced due to a biological vacuum. Furthermore, environmental conditions of the greenhouse can be manipulated to achieve BCA growth needs. Theoretically all these characteristics allow a better introduction, establishment and interaction of the BCA with plants in hydroponics than in soil (Paulitz and Bélanger 2001; Postma et al. 2009; Vallance et al. 2010). However, in practice, the effectiveness of BCA inoculation to control root pathogens can be highly variable in soilless systems (Postma et al. 2008; Vallance et al. 2010; Montagne et al. 2017). One explanation for this is that BCA selection is based on in vitro tests which are not representing real conditions and subsequently a weak adaptation of these microorganisms to the aquatic environment used in hydroponics or aquaponics (Postma et al. 2008; Vallance et al. 2010). To control plant pathogens and more especially those responsible for root rots, a selection and identification of microorganisms involved in aquatic systems which show suppressive activity against plant pathogens is needed. In soilless culture, several antagonistic microorganisms can be picked due to their biological cycle being similar to root pathogens or their ability to grow in aqueous conditions. Such is the case of nonpathogenic

Pythium and *Fusarium* species and bacteria, where *Pseudomonas*, *Bacillus* and *Lysobacter* are the genera most represented in the literature (Paulitz and Bélanger 2001; Khan et al. 2003; Chatterton et al. 2004; Folman et al. 2004; Sutton et al. 2006; Liu et al. 2007; Postma et al. 2008; Postma et al. 2009; Vallance et al. 2010; Sopher and Sutton 2011; Hultberg et al. 2011; Lee and Lee 2015; Martin and Loper 1999; Moruzzi et al. 2017; Thongkamngam and Jaenaksorn 2017). The direct addition of some microbial metabolites such as biosurfactants has also been studied (Stanghellini and Miller 1997; Nielsen et al. 2006; Nielsen et al. 2006). Although some microorganisms are efficient at controlling root pathogens, there are other problems that need to be overcome in order to produce a biopesticide. The main challenges are to determine the means of inoculation, the inoculum density, the product formulation (Montagne et al. 2017), the method for the production of sufficient quantity at low cost and the storage of the formulated product. Ecotoxicological studies on fish and living beneficial microorganisms in the system are also an important point. Another possibility that could be exploited is the use of a complex of antagonistic agents, as observed in suppressive soil techniques (Spadaro and Gullino 2005; Vallance et al. 2010). In fact, microorganisms can work in synergy or with complementary modes of action (ibid.). The addition of amendments could also enhance the BCA potential by acting as prebiotics.

The Role of Organic Matter in Biocontrol Activity in Aquaponic Systems

In Section, the suppressiveness of aquaponic systems was suggested. As stated before, the main hypothesis is related to the water recirculation as it is for hydroponic systems. However, a second hypothesis exists and this is linked to the presence of organic matter in the system. Organic matter that could drive a more balanced microbial ecosystem including antagonistic agents which is less suitable for plant pathogens (Rakocy 2012).

In aquaponics, organic matter comes from water supply, uneaten feeds, fish faeces, organic plant substrate, microbial activity, root exudates and plant residues (Waechter-Kristensen et al. 1997; Naylor et al. 1999; Waechter-Kristensen et al. 1999). In such a system, heterotrophic bacteria are organisms able to use organic matter as a carbon and energy source, generally in the form of carbohydrates, amino acids, peptides or lipids (Sharrer et al. 2005; Willey et al. 2008; Whipps 2001). In recirculated aquaculture (RAS), they are mainly localised in the biofilter and consume organic particles trapped in it (Leonard et al. 2000; Leonard et al. 2002). However, another source of organic carbon for heterotrophic bacteria is humic substances present as dissolved organic matter and responsible for the yellow-brownish coloration of the water (Takeda and Kiyono 1990 cited by Leonard et al. 2002; Hirayama et al. 1988). In the soil as well as in hydroponics, humic acids are known to stimulate plant growth and sustain the plant under abiotic stress conditions (Bohme 1999; du Jardin 2015). Proteins in the water can be used by plants as an alternative nitrogen source thus enhancing their growth and pathogen resistance (Adamczyk et al. 2010). In the recirculated water, the abundance of free-living heterotrophic bacteria is correlated with the amount of biologically available organic carbon and carbon-nitrogen ratio (C/N) (Leonard et al. 2000; Leonard et al. 2002; Michaud et al. 2006; Attramadal et al. 2012). In the biofilter, an increase in the C/N ratio increases the abundance of heterotrophic bacteria at the expense of the number of autotrophic bacteria responsible for the nitrification process (Michaud et al. 2006; Michaud et al. 2014). As implied, heterotrophic microorganisms can have a negative impact on the system because they compete with autotrophic bacteria (e.g. nitrifying bacteria) for space and oxygen. Some of them are plant or fish pathogens, or responsible for off-flavour in fish (Chang-Ho 1970; Funck-Jensen and Hockenhull 1983; Jones et al. 1991; Leonard et al. 2002; Nogueira et al. 2002; Michaud et al. 2006; Mukerji 2006; Whipps 2001; Rurangwa and Verdegem 2015). However, heterotrophic microorganisms involved in the system can also be positive (Whipps 2001; Mukerji 2006). Several studies using organic fertilizers or organic soilless media, in hydroponics, have shown interesting effects where the resident microbiota were able to control plant diseases (Montagne et al. 2015). All organic substrates have their own physico-chemical properties. Consequently, the characteristics of the media will influence microbial richness and functions. The choice of a specific plant media could therefore influence the microbial development so as to have a suppressive effect on pathogens (Montagne et al. 2015; Grunert et al. 2016; Montagne et al. 2017). Another possibility of pathogen suppression related to organic carbon is the use of organic amendments in hydroponics (Maher et al. 2008; Vallance et al. 2010). By adding composts in soilless media like it is common use in soil, suppressive effects are expected (Maher et al. 2008). Enhancing or maintaining a specific microorganism such as *Pseudomonas* population by adding some formulated carbon sources (e.g. nitrapyrin-based product) as reported by Pagliaccia et al. (2007) and Pagliaccia et al. (2008) is another possibility. The emergence of organic soilless culture also highlights the involvement of beneficial microorganisms against plant pathogens supported by the use of organic fertilizers. Fujiwara et al. (2013), Chinta et al. (2014), and Chinta et al. (2015) reported that fertilization with corn steep liquor helps to control *Fusarium oxysporum* f.sp. *lactucae* and *Botrytis cinerea* on lettuces and *Fusarium oxysporum* f.sp. *radicis-lycopersici* on tomato plants. And even if hardly advised for aquaponic use, 1 g/L of fish-based soluble fertilizer (Shinohara et al. 2011) suppresses bacterial wilt on tomato caused by *Ralstonia solanacearum* in hydroponics (Fujiwara et al. 2012).

Finally, though information about the impact of organic matter on plant protection in aquaponics is scarce, the various elements mentioned above show their potential capacity to promote a system-specific and plant pathogen-suppressive microbiota.

Conclusions and Future Considerations

This chapter aimed to give a first report of plant pathogens occurring in aquaponics, reviewing actual methods and future possibilities to control them. Each strategy has advantages and disadvantages and must be thoroughly designed to fit each case. However, at this time, curative methods in coupled aquaponic systems are still limited and new perspectives of control must be found. Fortunately, suppressiveness in terms of aquaponic systems could be considered, as already observed in hydroponics (e.g. in plant media, water, and slow filters). In addition, the presence of organic matter in the system is an encouraging factor when compared to soilless culture systems making use of organic fertilisers, organic plant media or organic amendments.

For the future, it seems important to investigate this suppressive action followed by identification and characterization of the responsible microbes or microbe consortia. Based on the results, several strategies could be envisaged to enhance the capacity of plants to resist pathogens. The *first* is biological control by conservation, which means favouring beneficial microorganisms by manipulating and managing water composition (e.g. C/N ratio, nutrients and gases) and parameters (e.g. pH and temperature). But identification of these influencing factors needs to be realized first. This management of autotrophic and heterotrophic bacteria is also of key importance to sustain good nitrification and keep healthy fish. The *second* strategy is augmentative biological control by additional release of beneficial microorganisms already present in the system in large numbers (inundative method) or in small numbers but repeated in time (inoculation method). But prior identification and multiplication of an aquaponic BCA should be achieved. The *third* strategy is importation, i.e. introducing a new microorganism normally not present into the system. In this case, selection of a microorganism adapted and safe for aquaponic environment is needed. For the two last strategies, the site of inoculation in the system must be considered depending on the aim desired. Sites where microbial activity could be enhanced are the recirculated water, the rhizosphere (plant media included), the biofilter (such as in slow sand filters where BCA addition is already tested) and the phyllosphere (i.e. aerial plant part). Whatever the strategy, the ultimate goal should be to lead the microbial communities to provide a stable, ecologically balanced microbial environment allowing good production of both plant and fish.

To conclude, following the requirements of integrated plant pest management (IPM) is a necessity to correctly manage the system and avoid development and spread of plant diseases (Bittsanszky et al. 2015; Nemethy et al. 2016). The principle of IPM is to apply chemical pesticides or other agents as a last resort when economic injury level is reached. Consequently, control of pathogens will need to be firstly based on physical and biological methods (described above), their combination and an efficient detection and monitoring of the disease (European Parliament 2009).

References

Adamczyk B, Smolander A, Kitunen V, Godlewski M (2010) Proteins as nitrogen source for plants: a short story about exudation of proteases by plant roots. Plant Signal Behav 5:817–819. https://doi.org/10.4161/psb.5.7.11699

Albajes R, Lodovica Gullino M, Van Lenteren JC, Elad Y (2002) Integrated pest and disease management in greenhouse crops. Kluwer Academic Publishers. https://doi.org/10.1017/CBO9781107415324.004

Alhussaen K (2006) Pythium and phytophthora associated with root disease of hydroponic lettuce. University of Technologie Sydney Faculty of Science. https://opus.lib.uts.edu.au/handle/10453/36864

Anderson TS, de Villiers D, Timmons MB (2017) Growth and tissue elemental composition response of Spinach (Spinacia oleracea) to hydroponic and aquaponic water quality conditions. Horticulturae 3:32. https://doi.org/10.3390/horticulturae3020032

Arras G, Arru S (1997) Mechanism of action of some microbial antagonists against fungal pathogens. Ann Microbiol Enzimol 47:97–120

Attramadal KJK, Salvesen I, Xue R, Øie G, Størseth TR, Vadstein O, Olsen Y (2012) Recirculation as a possible microbial control strategy in the production of marine larvae. Aquac Eng 46:27–39. https://doi.org/10.1016/j.aquaeng.2011.10.003

Bartelme RP, Oyserman BO, Blom JE, Sepulveda-Villet OJ, Newton RJ (2018) Stripping away the soil: plant growth promoting microbiology opportunities in aquaponics. Front Microbiol 9:8. https://doi.org/10.3389/fmicb.2018.00008

Beneduzi A, Ambrosini A, Passaglia LMP (2012) Plant growth-promoting rhizobacteria (PGPR): their potential as antagonists and biocontrol agents. Genet Mol Biol 35(4):1044–1051

Bittsanszky A, Gyulai G, Junge R, Schmautz Z, Komives T, Has CAR, Otto H (2015) Plant protection in ecocycle-based agricultural systems : aquaponics as an example. In: International Plant Protection Congress (IPPC), Berlin, Germany, pp 2–3. https://doi.org/10.13140/RG.2.1.4458.0321

Bohme M (1999) Effects of Lactate, Humate and Bacillus subtilis on the growth of tomato plants in hydroponic systems. In: International symposium on growing media and hydroponics. Acta Horticulturae, pp 231–239

Brand T (2001) Importance and characterization of the biological component in slow filters. Acta Hortic 554:313–321

Calvo-bado LA, Pettitt TR, Parsons N, Petch GM, Morgan JAW, Whipps JM (2003) Spatial and Temporal Analysis of the Microbial Community in Slow Sand Filters Used for Treating Horticultural Irrigation Water. Appl Enviromental Microbiol 69:2116–2125. https://doi.org/10.1128/AEM.69.4.2116

Campbell R (1989) Biological control of microbial plant pathogens. Cambridge University Press, Cambridge

Chang-Ho Y (1970) The effect of pea root exudate on the germination of Pythium aphanidermatum zoospore cysts. Can J Bot 48:1501–1514

Chatterton S, Sutton JC, Boland GJ (2004) Timing Pseudomonas chlororaphis applications to control Pythium aphanidermatum, Pythium dissotocum, and root rot in hydroponic peppers. Biol Control 30:360–373. https://doi.org/10.1016/j.biocontrol.2003.11.001

Cherif M, Tirilly Y, Belanger RR (1997) Effect of oxygen concentration on plant growth, lipid peroxidation, and receptivity of tomato roots to Pythium under hydroponic conditions. Eur J Plant Pathol 103:255–264

Chinta YD, Kano K, Widiastuti A, Fukahori M, Kawasaki S, Eguchi Y, Misu H, Odani H, Zhou S, Narisawa K, Fujiwara K, Shinohara M, Sato T (2014) Effect of corn steep liquor on lettuce root rot (Fusarium oxysporum f.sp. lactucae) in hydroponic cultures. J Sci Food Agric 94:2317–2323. https://doi.org/10.1002/jsfa.6561

Chinta YD, Eguchi Y, Widiastuti A, Shinohara M, Sato T (2015) Organic hydroponics induces systemic resistance against the air-borne pathogen, *Botrytis cinerea* (gray mould). J Plant Interact 10:243–251. https://doi.org/10.1080/17429145.2015.1068959

Colhoun J (1973) Effects of environmental factors on plant disease. Annu Rev Phytopathol 11:343–364

da Silva Cerozi B, Fitzsimmons K (2016) Use of Bacillus spp. to enhance phosphorus availability and serve as a plant growth promoter in aquaponics systems. Sci Hortic (Amsterdam) 211:277–282. https://doi.org/10.1016/j.scienta.2016.09.005

Delaide B, Goddek S, Gott J, Soyeurt H, Jijakli HM (2016) Lettuce (Lactuca sativa L. var. Sucrine) growth performance in complemented aquaponic solution outperforms hydroponics. Water 8:1–11. https://doi.org/10.3390/w8100467

Déniel F, Rey P, Chérif M, Guillou A, Tirilly Y (2004) Indigenous bacteria with antagonistic and plant-growth-promoting activities improve slow-filtration efficiency in soilless cultivation. Can J Microbiol 50:499–508. https://doi.org/10.1139/w04-034

Dordas C (2008) Role of nutrients in controlling plant diseases in sustainable agriculture: a review. Agron Sustain Dev 28:33–46. https://doi.org/10.1051/agro:2007051

du Jardin P (2015) Plant biostimulants: definition, concept, main categories and regulation. Sci Hortic (Amsterdam) 196:3–14. https://doi.org/10.1016/j.scienta.2015.09.021

Ehret DL, Bogdanoff C, Utkhede R, Lévesque A, Menzies JG, Ng K, Portree J (1999) Disease control with slow filtration for greenhouse crops grown in recirculation. Pacific Agri-Food Res Cent Cent Tech Rep 155:37

Ehret DL, Alsanius B, Wohanka W, Menzies JG, Utkhede R (2001) Disinfestation of recirculating nutrient solutions in greenhouse horticulture. Agronomie 21:323–339. https://doi.org/10.1051/agro:2001127

European Parliament (2009) Directive 2009/128/EC of the European Parliament and the Council of 21 October 2009 establishing a framework for Community action to achieve the sustainable use of pesticides. October 309, pp 71–86. https://doi.org/10.3000/17252555.L_2009.309

Evenhuis B, Nijhuis E, Lamers J, Verhoeven J, Postma J (2014) Alternative methods to control phytophthora cactorum in strawberry cultivated in soilless growing media, Acta Hortic 1044:337–342. https://doi.org/10.17660/ActaHortic.2014.1044.44

FAO (2008) Good agricultural practices [WWW Document]. http://www.fao.org/prods/gap/. Accessed 27 Feb 2018

Folman LB, De Klein MJEM, Postma J, Van Veen JA (2004) Production of antifungal compounds by Lysobacter enzymogenes isolate 3.1T8 under different conditions in relation to its efficacy as a biocontrol agent of Pythium aphanidermatum in cucumber. Biol. Control 31:145–154. https://doi.org/10.1016/j.biocontrol.2004.03.008

Fujiwara K, Aoyama C, Takano M, Shinohara M (2012) Suppression of Ralstonia solanacearum bacterial wilt disease by an organic hydroponic system. J Gen Plant Pathol. https://doi.org/10.1007/s10327-012-0371-0

Fujiwara K, Iida Y, Iwai T, Aoyama C, Inukai R, Ando A, Ogawa J, Ohnishi J, Terami F, Takano M, Shinohara M (2013) The rhizosphere microbial community in a multiple parallel mineralization system suppresses the pathogenic fungus Fusarium oxysporum. Microbiologyopen 2:997–1009. https://doi.org/10.1002/mbo3.140

Funck-Jensen D, Hockenhull J (1983) Is damping-off, caused by Pythium, less of a problem in hydroponics than in traditional growing systems? Acta Hortic 133:137–145

Furtner B, Bergstrand K, Brand T (2007) Abiotic and biotic factors in slow filters integrated to closed hydroponic systems. Eur J Hortic Sci 72:104–112

Ganeshan G, Kumar AM (2005) Pseudomonas fluorescens, a potential bacterial antagonist to control plant diseases. J Plant Interact 1:123–134. https://doi.org/10.1080/17429140600907043

Geary B, Clark J, Hopkins BG, Jolley VD (2015) Deficient, adequate and excess nitrogen levels established in hydroponics for biotic and abiotic stress-interaction studies in potato. J Plant Nutr 38:41–50. https://doi.org/10.1080/01904167.2014.912323

Goddek S, Vermeulen T (2018) Comparison of Lactuca sativa growth performance in rainwater and RAS-water-based hydroponic nutrient solutions. Aquac Int 10. https://doi.org/10.1007/s10499-018-0293-8

Goddek S, Delaide B, Mankasingh U, Vala Ragnarsdottir K, Jijakli H, Thorarinsdottir R (2015) Challenges of sustainable and commercial aquaponics. Sustainability 7:4199–4224. https://doi.org/10.3390/su7044199

Goddek S, Espinal CA, Delaide B, Jijakli MH, Schmautz Z, Wuertz S, Keesman KJ (2016) Navigating towards decoupled aquaponic systems: a system dynamics design approach. Water (Switzerland) 8:1–29. https://doi.org/10.3390/W8070303

Gonçalves AA, Gagnon GA (2011) Ozone application in recirculating aquaculture system: an overview. Ozone Sci Eng 33:345–367. https://doi.org/10.1080/01919512.2011.604595

Gravel V, Dorais M, Dey D, Vandenberg G (2015) Fish effluents promote root growth and suppress fungal diseases in tomato transplants. Can J Plant Sci 95:427–436

Grunert O, Hernandez-Sanabria E, Vilchez-Vargas R, Jauregui R, Pieper DH, Perneel M, Van Labeke M-C, Reheul D, Boon N (2016) Mineral and organic growing media have distinct community structure, stability and functionality in soilless culture systems. Sci. Rep. 6:18837. https://doi.org/10.1038/srep18837

Haarhoff J, Cleasby JL (1991) Biological and physical mechanisms in slow sand filtration. In: Logsdon GS (ed) Slow sand filtration, vol 1. American Society of Civil Engineers, New York, pp 19–68

Haas D, Défago G (2005) Biological control of soil-borne pathogens by fluorescent pseudomonads. Nat Rev Microbiol 3:307–319. https://doi.org/10.1038/nrmicro1129

Hirayama K, Mizuma H, Mizue Y (1988) The accumulation of dissolved organic substances in closed recirculation culture systems. Aquac Eng 7:73–87. https://doi.org/10.1016/0144-8609(88)90006-4

Hong CX, Moorman GW (2005) Plant pathogens in irrigation water: challenges and opportunities. CRC Crit Rev Plant Sci 24:189–208. https://doi.org/10.1080/07352680591005838

Hultberg M, Holmkvist A, Alsanius B (2011) Strategies for administration of biosurfactant-producing pseudomonads for biocontrol in closed hydroponic systems. Crop Prot 30:995–999. https://doi.org/10.1016/j.cropro.2011.04.012

James, Becker JO (2007) Identifying microorganisms involved in specific pathogen suppression in soil. Annu Rev Phytopathol 45:153–172. https://doi.org/10.1146/annurev.phyto.45.062806.094354

Jarvis WR (1992) Managing disease in greenhouse crops. The American Phytopathological Society, St. Paul

Jones SW, Donaldson SP, Deacon JW (1991) Behaviour of zoospores and zoospore cysts in relation to root infection by *Pythium aphanidermatum*. New Phytol 117:289–301. https://doi.org/10.1111/j.1469-8137.1991.tb04910.x

Khalil S, Alsanius BW (2001) Dynamics of the indigenous microflora inhabiting the root zone and the nutrient solution of tomato in a commercial closed greenhouse system. Gartenbauwissenschaft 66:188–198

Khalil S, Alsanius WB (2011) Effect of growing medium water content on the biological control of root pathogens in a closed soilless system. J Hortic Sci Biotechnol 86:298–304. https://doi.org/10.1080/14620316.2011.11512764

Khalil S, Hultberg M, Alsanius BW (2009) Effects of growing medium on the interactions between biocontrol agents and tomato root pathogens in a closed hydroponic system. J Hortic Sci Biotechnol 84:489–494. https://doi.org/10.1080/14620316.2009.11512553

Khan A, Sutton JC, Grodzinski B (2003) Effects of Pseudomonas chlororaphis on Pythium aphanidermatum and root rot in peppers grown in small-scale hydroponic troughs. Biocontrol Sci Technol 13:615–630. https://doi.org/10.1080/0958315031000151783

Koohakan P, Ikeda H, Jeanaksorn T, Tojo M, Kusakari S-I, Okada K, Sato S (2004) Evaluation of the indigenous microorganisms in soilless culture: occurrence and quantitative characteristics in the different growing systems. Sci Hortic (Amsterdam) 101:179–188. https://doi.org/10.1016/j.scienta.2003.09.012

Lee S, Lee J (2015) Beneficial bacteria and fungi in hydroponic systems: types and characteristics of hydroponic food production methods. Sci Hortic (Amsterdam) 195:206–215. https://doi.org/10.1016/j.scienta.2015.09.011

Leonard N, Blancheton JP, Guiraud JP (2000) Populations of heterotrophic bacteria in an experimental recirculating aquaculture system. Aquac Eng 22:109–120

Leonard N, Guiraud JP, Gasset E, Cailleres JP, Blancheton JP (2002) Bacteria and nutrients – nitrogen and carbon – in a recirculating system for sea bass production. Aquac Eng 26:111–127

Lepoivre P (2003) Phytopathologie, 1st editio. ed. Les Presses Agronomiques de Gembloux Bruxelles: De Boeck

Liu W, Sutton JC, Grodzinski B, Kloepper JW, Reddy MS (2007) Biological control of pythium root rot of chrysanthemum in small-scale hydroponic units. Phytoparasitica 35:159–178. https://doi.org/10.1007/BF02981111

Love DC, Fry JP, Li X, Hill ES, Genello L, Semmens K, Thompson RE (2015) Commercial aquaponics production and profitability: Findings from an international survey. Aquaculture 435:67–74. https://doi.org/10.1016/j.aquaculture.2014.09.023

Maher M, Prasad M, Raviv M (2008) Ch 11 – organic soilless media components. In: Soilless culture: theory and practice. Elsevier B.V., Amsterdam, pp 459–504. https://doi.org/10.1016/B978-044452975-6.50013-7

Mangmang JS, Deaker R, Rogers G (2014) Response of lettuce seedlings fertilized with fish effluent to Azospirillum brasilense inoculation. Biol Agric Hortic 31:61–71. https://doi.org/10.1080/01448765.2014.972982

Mangmang JS, Deaker R, Rogers G (2015a) Inoculation effect of Azospirillum brasilense on basil grown under aquaponics production system. Org Agric 6:65–74. https://doi.org/10.1007/s13165-015-0115-5

Mangmang JS, Deaker R, Rogers G (2015b) Maximizing fish effluent utilization for vegetable seedling production by Azospirillum Brasilense. Procedia Environ Sci 29:179. https://doi.org/10.1016/j.proenv.2015.07.248

Mangmang JS, Deaker R, Rogers G (2015c) Response of Cucumber Seedlings Fertilized with Fish Effluent to Azospirillum brasilense. Int J Veg Sci 5260:150409121518007. https://doi.org/10.1080/19315260.2014.967433

Martin FN, Loper JE (1999) Soilborne plant diseases caused by Pythium spp.: ecology, epidemiology, and prospects for biological control soilborne plant diseases caused by Pythium spp.: ecology, epidemiology, and prospects for biological control. CRC Crit Rev Plant Sci 18:11–181

McMurty MR, Nelson PV, Sanders DC, Hodges L (1990) Sand culture of vegetables using recirculated aquacultural effluents. Appl Agric Res 5:280–284

McPherson GM, Harriman MR, Pattison D (1995) The potential for spread of root diseases in recirculating hydroponic systems and their control with disinfection. Med Fac Landbouww Univ Gent 60:371–379

Michaud L, Blancheton JP, Bruni V, Piedrahita R (2006) Effect of particulate organic carbon on heterotrophic bacterial populations and nitrification efficiency in biological filters. Aquac Eng 34:224–233. https://doi.org/10.1016/j.aquaeng.2005.07.005

Michaud L, Giudice AL, Interdonato F, Triplet S, Ying L, Blancheton JP (2014) C/N ratio-induced structural shift of bacterial communities inside lab-scale aquaculture biofilters. Aquac Eng 58:77–87. https://doi.org/10.1016/j.aquaeng.2013.11.002

Mitchell CE, Reich PB, Tilman D, Groth JV (2003) Effects of elevated CO_2, nitrogen deposition, and decreased species diversity on foliar fungal plant disease. Glob Chang Biol 9, 438–451. https://doi.org/10.1046/j.1365-2486.2003.00602.x

Monsees H, Kloas W, Wuertz S (2017) Decoupled systems on trial: Eliminating bottlenecks to improve aquaponic processes. PLoS One 12:1–18. https://doi.org/10.1371/journal.pone.0183056

Montagne V, Charpentier S, Cannavo P, Capiaux H, Grosbellet C, Lebeau T (2015) Structure and activity of spontaneous fungal communities in organic substrates used for soilless crops. Sci Hortic (Amsterdam) 192:148–157. https://doi.org/10.1016/j.scienta.2015.06.011

Montagne V, Capiaux H, Barret M, Cannavo P, Charpentier S, Grosbellet C, Lebeau T (2017) Bacterial and fungal communities vary with the type of organic substrate: implications for biocontrol of soilless crops. Environ Chem Lett 15:537–545. https://doi.org/10.1007/s10311-017-0628-0

Moruzzi S, Firrao G, Polano C, Borselli S, Loschi A, Ermacora P, Loi N, Martini M (2017) Genomic-assisted characterisation of Pseudomonas sp. strain Pf4, a potential biocontrol agent in hydroponics. Biocontrol Sci Technol 27:969–991. https://doi.org/10.1080/09583157.2017.1368454

Mukerji KG (2006) Microbial Activity in the Rhizosphere. Springer/GmbH & Co, Dordrecht/Berlin and Heidelberg. https://doi.org/10.1017/CBO9781107415324.004

Munguia-Fragozo P, Alatorre-Jacome O, Rico-Garcia E, Torres-Pacheco I, Cruz-Hernandez A, Ocampo-Velazquez RV, Garcia-Trejo JF, Guevara-Gonzalez RG (2015) Perspective for Aquaponic Systems: (Omic) Technologies for Microbial Community Analysis. Biomed Res Int 2015:10. https://doi.org/10.1155/2015/480386

Narayanasamy P (2013) Biological management of diseases of crops: volume 1: characteristics of biological control agents, Biological management of diseases of crops. Springer, Dordrecht. https://doi.org/10.1007/978-94-007-6380-7_1

Naylor SJ, Moccia RD, Durant GM (1999) The chemical composition of settleable solid fish waste (manure) from commercial rainbow trout farms in Ontario. Canada N Am J Aquac 61:21–26. https://doi.org/10.1577/1548-8454(1999)061<0021:TCCOSS>2.0.CO;2

Nemethy S, Bittsanszky A, Schmautz Z, Junge R, Komives T (2016) Protecting plants from pests and diseases in aquaponic systems. In: Ecological footprint in Central Europe. The University College of Tourism and Ecology Press, Sucha Beskidzka, pp 1–8

Nielsen CJ, Ferrin DM, Stanghellini ME (2006) Efficacy of biosurfactants in the management of Phytophthora capsici on pepper in recirculating hydroponic systems. Can J Plant Pathol 28:450–460. https://doi.org/10.1080/07060660609507319

Nogueira R, Melo LF, Purkhold U, Wuertz S, Wagner M (2002) Nitrifying and heterotrophic population dynamics in biofilm reactors: effects of hydraulic retention time and the presence of organic carbon. Water Res 36:469–481

Pagliaccia D, Ferrin D, Stanghellini ME (2007) Chemo-biological suppression of root-infecting zoosporic pathogens in recirculating hydroponic systems. Plant Soil 299:163–179. https://doi.org/10.1007/s11104-007-9373-7

Pagliaccia D, Merhaut D, Colao MC, Ruzzi M, Saccardo F, Stanghellini ME (2008) Selective enhancement of the fluorescent pseudomonad population after amending the recirculating nutrient solution of hydroponically grown plants with a nitrogen stabilizer. Microb Ecol 56:538–554. https://doi.org/10.1007/s00248-008-9373-z

Pantanella E, Cardarelli M, Colla G, Rea E, Marcucci A (2010) Aquaponics vs. hydroponics: production and quality of lettuce crop, pp 887–893. https://doi.org/10.17660/ActaHortic.2012.927.109

Pantanella E, Cardarelli M, Di Mattia E, Colla G (2015) Aquaponics and food safety: effects of UV sterilization on total coliforms and lettuce production. In: Conference and exhibition on soilless culture, pp 71–76

Parvatha Reddy P (2016) Sustainable crop protection under protected cultivation. Springer. https://doi.org/DOI, Springer. https://doi.org/10.1007/978-981-287-952-3_7

Paulitz TC, Bélanger RR (2001) Biological control in greenhouse systems. Annu Rev Phytopathol 39:103–133

Pérez-García A, Romero D, de Vicente A (2011) Plant protection and growth stimulation by microorganisms: biotechnological applications of Bacilli in agriculture. Curr Opin Biotechnol 22(2):187–193. https://doi.org/10.1016/j.copbio.2010.12.003

Postma J, van Os E, Bonants PJM (2008) Ch 10 – pathogen detection and management strategies in soilless plant growing system. In: Soilless culture: theory and practice. Elsevier B.V., Amsterdam, pp. 425–457. https://doi.org/10.1016/B978-0-444-52975-6.50012-5

Postma J, Stevens LH, Wiegers GL, Davelaar E, Nijhuis EH (2009) Biological control of Pythium aphanidermatum in cucumber with a combined application of Lysobacter enzymogenes strain 3.1T8 and chitosan. Biol Control 48:301–309. https://doi.org/10.1016/j.biocontrol.2008.11.006

Rakocy J (2007) Ten guidelines for aquaponic systems. Aquaponics J 46:14–17

Rakocy JE (2012) Aquaponics – integrating fish and plant culture. In: Tidwell JH (ed) Aquaculture production systems. Wiley, New York, pp 343–386

Rakocy JE, Maseer PM, Losordo TM (2006) Recirculating aquaculture tank production systems: Aquaponics – integrated fish and plant culture. South Reg Aquac Cent Publication No. 454, 16pp. http://srac.tamu.edu/getfile.cfm?pubid=105

Renault D, Déniel F, Benizri E, Sohier D, Barbier G, Rey P (2007) Characterization of Bacillus and Pseudomonas strains with suppressive traits isolated from tomato hydroponic-slow filtration unit. Can J Microbiol 53:784–797. https://doi.org/10.1139/W07-046

Renault D, Vallance J, Déniel F, Wery N, Godon JJ, Barbier G, Rey P (2012) Diversity of bacterial communities that colonize the filter units used for controlling plant pathogens in soilless cultures. Microb Ecol 63:170–187. https://doi.org/10.1007/s00248-011-9961-1

Resh HM (2013) Hydroponic food production : a definitive guidebook for the advanced home gardener and the commercial hydroponic grower, 7th edn. CRC Press, Boca Raton

Rosberg AK (2014) Dynamics of root microorganisms in closed hydroponic cropping systems. Department of Biosystems and Technology, Swedish University of Agricultural Science, Alnarp

Rurangwa E, Verdegem MCJ (2015) Microorganisms in recirculating aquaculture systems and their management. Rev Aquac 7:117–130. https://doi.org/10.1111/raq.12057

Saha S, Monroe A, Day MR (2016) Growth, yield, plant quality and nutrition of basil (*Ocimum basilicum* L.) under soilless agricultural systems. Ann Agric Sci 61:181–186. https://doi.org/10.1016/j.aoas.2016.10.001

Schmautz Z, Graber A, Jaenicke S, Goesmann A, Junge R, Smits THM (2017) Microbial diversity in different compartments of an aquaponics system. Arch Microbiol 1–8. https://doi.org/10.1007/s00203-016-1334-1

Schreier HJ, Mirzoyan N, Saito K (2010) Microbial diversity of biological filters in recirculating aquaculture systems. Curr Opin Biotechnol 21:318–325. https://doi.org/10.1016/j.copbio.2010.03.011

Sharrer MJ, Summerfelt ST, Bullock GL, Gleason LE, Taeuber J (2005) Inactivation of bacteria using ultraviolet irradiation in a recirculating salmonid culture system. Aquac Eng 33:135–149. https://doi.org/10.1016/j.aquaeng.2004.12.001

Shinohara M, Aoyama C, Fujiwara K, Watanabe A, Ohmori H, Uehara Y, Takano M (2011) Microbial mineralization of organic nitrogen into nitrate to allow the use of organic fertilizer in hydroponics. Soil Sci Plant Nutr 57:190–203. https://doi.org/10.1080/00380768.2011.554223

Sirakov I, Lutz M, Graber A, Mathis A, Staykov Y (2016) Potential for combined biocontrol activity against fungal fish and plant pathogens by bacterial isolates from a model aquaponic system. Water 8:1–7. https://doi.org/10.3390/w8110518

Snoeijers SS, Alejandro P (2000) The effect of nitrogen on disease development and gene expression in bacterial and fungal plant pathogens. Eur J Plant Pathol 106:493–506

Somerville C, Cohen M, Pantanella E, Stankus A, Lovatelli A (2014) Small-scale aquaponic food production – integrated fish and plant farming. FAO, Rome

Sopher CR, Sutton JC (2011) Quantitative relationships of Pseudomonas chlororaphis 63-28 to Pythium root rot and growth in hydroponic peppers. Trop Plant Pathol 36:214–224. https://doi.org/10.1590/S1982-56762011000400002

Spadaro D, Gullino ML (2005) Improving the efficacy of biocontrol agents against soilborne pathogens. Crop Prot 24:601–613. https://doi.org/10.1016/j.cropro.2004.11.003

Stanghellini ME, Miller RM (1997) Their identity and potencial efficacy in the biological control of zoosporic plant pathogens. Plant Dis 81:4–12

Stanghellini ME, Rasmussen SL (1994) Hydroponics: a solution for zoosporic pathogens. Plant Dis 78:1129–1138

Stouvenakers G, Sébastien M, Haissam JM (2017) Biocontrol properties of recirculating aquaculture water against hydroponic root pathogens. Oral presentation at Aquaculture Europe 2017 meeting, Dubrovnik, Croatia

Sugita H, Nakamura H, Shimada T (2005) Microbial communities associated with filter materials in recirculating aquaculture systems of freshwater fish. Aquaculture 243:403–409. https://doi.org/10.1016/j.aquaculture.2004.09.028

Sutton JC, Sopher CR, Owen-Going TN, Liu W, Grodzinski B, Hall JC, Benchimol RL (2006) Etiology and epidemiology of Pythium root rot in hydroponic crops: current knowledge and perspectives. Summa Phytopathol 32:307–321. https://doi.org/10.1590/S0100-54052006000400001

Takeda S, Kiyono M (1990) The characterisation of yellow substances accumulated in a closed recirculating system for fish culture. In: Proceedings of the second Asian fisheries forum, pp 129–132

Tal Y, Watts JEM, Schreier SB, Sowers KR, Schreier HJ, Schreier HJ (2003) Characterization of the microbial community and nitrogen transformation processes associated with moving bed bioreactors in a closed recirculated mariculture system. Aquaculture 215:187–202

Thongkamngam T, Jaenaksorn T (2017) Fusarium oxysporum (F221-B) as biocontrol agent against plant pathogenic fungi in vitro and in hydroponics. Plant Prot Sci 53:85–95. https://doi.org/10.17221/59/2016-PPS

Timmons MB, Ebeling JM (2010) Recirculating aquaculture, 2nd edn. NRAC Publication, Ithaca

Tu JC, Papadopoulos AP, Hao X, Zheng J (1999) The relationship of a pythium root rot and rhizosphere microorganisms in a closed circulating and an open system in stone wool culture of tomato. Acta Hort (ISHS) 481:577–583

Vallance J, Déniel F, Le Floch G, Guérin-Dubrana L, Blancard D, Rey P (2010) Pathogenic and beneficial microorganisms in soilless cultures. Agron Sustain Dev 31:191–203. https://doi.org/10.1051/agro/2010018

Van Der Gaag DJ, Wever G (2005) Conduciveness of different soilless growing media to Pythium root and crown rot of cucumber under near-commercial conditions. Eur J Plant Pathol 112:31–41. https://doi.org/10.1007/s10658-005-1049-7

Van Os EA (2009) Comparison of some chemical and non-chemical treatments to disinfect a recirculating nutrient solution. Acta Hortic 843:229–234

Van Os EA, Amsing JJ, Van Kuik AJ, Willers H (1999) Slow sand filtration: a potential method for the elimination of pathogens and nematodes in recirculating nutrient solutions from glasshouse-grown crops. Acta Hortic 481:519–526

van Os EA, Bruins M, Wohanka W, Seidel R (2001) Slow filtration: a technique to minimise the risks of spreading root-infecting pathogens in closed hydroponic systems. In: International symposium on protected cultivation in mild winter climates: current trends for sustainable techniques, pp 495–502

Veresoglou SD, Barto EK, Menexes G, Rillig MC (2013) Fertilization affects severity of disease caused by fungal plant pathogens. Plant Pathol 62:961–969. https://doi.org/10.1111/ppa.12014

Verma S, Daverey A, Sharma A (2017) Slow sand filtration for water and wastewater treatment – a review. Environ Technol Rev 6:47–58. https://doi.org/10.1080/21622515.2016.1278278

Villarroel M, Junge R, Komives T, König B, Plaza I, Bittsánszky A, Joly A (2016) Survey of aquaponics in Europe. Water (Switzerland) 8:3–9. https://doi.org/10.3390/w8100468

Waechter-Kristensen B, Sundin P, Gertsson UE, Hultberg M, Khalil S, Jensen P, Berkelmann-Loehnertz B, Wohanka W (1997) Management of microbial factors in the rhizosphere and nutrient solution of hydroponically grown tomato. Acta Hortic. https://doi.org/10.17660/ActaHortic.1997.450.40

Waechter-Kristensen B, Caspersen S, Adalsteinsson S, Sundin P, Jensén P (1999) Organic compounds and micro-organisms in closed, hydroponic culture: occurrence and effects on plant growth and mineral nutrition. Acta Hortic 481:197–204

Whipps JM (2001) Microbial interactions and biocontrol in the rhizosphere. J Exp Bot 52:487–511. https://doi.org/10.1093/jexbot/52.suppl_1.487

Wielgosz ZJ, Anderson TS, Timmons MB (2017) Microbial effects on the production of aquaponically Grown Lettuce. Horticulturae 3:46. https://doi.org/10.3390/horticulturae3030046

Willey JM, Sherwood LM, Woolverton CJ (2008) Prescott, Harley, & Klein's micobiology, 7th edn. McGrawHill Higher Education, New York

Wohanka W (1995) Disinfection of recirculating nutrient solutions by slow sand filtration. Acta Hortic. https://doi.org/10.17660/ActaHortic.1995.382.28

Zou Y, Hu Z, Zhang J, Xie H, Liang S, Wang J, Yan R (2016) Attempts to improve nitrogen utilization efficiency of aquaponics through nitrifies addition and filler gradation. Environ Sci Pollut Res 23:6671–6679. https://doi.org/10.1007/s11356-015-5898-0

8

Smarthoods: Aquaponics Integrated Microgrids

Florijn de Graaf and Simon Goddek

Abstract With the pressure to transition towards a fully renewable energy system increasing, a new type of power system architecture is emerging: the microgrid. A microgrid integrates a multitude of decentralised renewable energy technologies using smart energy management systems, in order to efficiently balance the local production and consumption of renewable energy, resulting in a high degree of flexibility and resilience. Generally, the performance of a microgrid increases with the number of technologies present, although it remains difficult to create a fully autonomous microgrid within economic reason (de Graaf F, New strategies for smart integrated decentralised energy systems, 2018). In order to improve the self-sufficiency and flexibility of these microgrids, this research proposes integrating a neighbourhood microgrid with an urban agriculture facility that houses a decoupled multi-loop aquaponics facility. This new concept is called *Smarthood*, where all Food–Water–Energy flows are circularly connected. In doing so, the performance of the microgrid greatly improves, due to the high flexibility present within the thermal mass, pumps and lighting systems. As a result, it is possible to achieve 95.38% power and 100% heat self-sufficiency. This result is promising, as it could pave the way towards realising these fully circular, decentralised Food–Water–Energy systems.

Keywords Smart grids · Urban agriculture · Microgrids · Blockchains · Self-sufficiency · Energy system modelling

---

F. de Graaf (✉)
Spectral, Amsterdam, The Netherlands
e-mail: florijn@spectral.energy

S. Goddek
Mathematical and Statistical Methods (Biometris), Wageningen University, Wageningen, The Netherlands
e-mail: simon.goddek@wur.nl; simon@goddek.nl

Introduction

Switching towards a fully sustainable energy system will partly require switching from a centralised generation and distribution system, towards a decentralised system, due to the rise of decentralised energy generation technologies using wind and rooftop solar radiation. In addition, integrating the heat and transport sectors into the electricity system will lead to a very significant increase in peak demand. These developments require massive and costly adaptations to the energy infrastructure, while the utilisation of existing production assets is expected to drop from 55% to 35% by 2035 (Strbac et al. 2015). This poses a major challenge, but also an opportunity: if the energy flows can be balanced locally in *microgrids*, the demand for expensive infrastructure upgrade can be minimised, while providing extra stability to the main grid. For these reasons, 'microgrids have been identified as a key component of the Smart Grid for improving power reliability and quality, increasing system energy efficiency' (Strbac et al. 2015).

Microgrids can provide much-needed resilience and flexibility, and are therefore likely to play an important role in the energy system of the future. It is estimated that by 2050, over half of EU households will be generating their own electricity (Pudjianto et al. 2007). Unlocking flexible resources within microgrids is therefore needed in order to balance the intermittent renewable energy generation.

Urban agriculture systems, such as aquaponics (dos Santos 2016), can provide this much-needed energy flexibility (Goddek and Körner 2019; Yogev et al. 2016). Plants can grow within a wide range of external conditions, since they are used to doing so in nature. The same applies to fish in an aquaculture system, which can thrive in a broad temperature range. These flexible operating conditions allow for a buffering effect on energy input requirements, which create a large degree of flexibility within the system. The high thermal mass embodied by the aquaculture system allows for vast amounts of heat to be stored within the system. The lights can be turned on and off depending on the abundance of electricity, allowing for excess electricity generation to essentially be curtailed by turning it into valuable biomass. Pumps can be operated in synchronicity with peak power generation times (e.g. noon) to limit net peak power (peak shaving). Optimal distillation units (Chapter) also have a very flexible heat demand and can be turned off as soon as there is an oversupply of heat or electricity (i.e. the heat pump would then convert electric energy into thermal energy). All these aspects make aquaponic systems well-suited to provide flexibility to a microgrid.

Next to providing flexibility in consumption, a multi-loop aquaponics system can be further integrated to also provide flexibility in production. Biogas is produced as a byproduct from the UASB in the aquaponic facility. This biogas can be combusted in order to produce both heat and power, by incorporating a micro-CHP in the microgrid. Integrating aquaponic systems within microgrids can therefore enhance energy flexibility both on the demand and supply sides.

The Smarthoods Concept

To unlock the full potential of the Food–Water–Energy nexus with respect to decentralised microgrids, a fully integrated approach focuses not only on energy (microgrid) and food (aquaponics) but also on utilising the local water cycle. The integration of various water systems (such as rainwater collection, storage and wastewater treatment) within aquaponic-integrated microgrids yields the biggest potential for efficiency, resilience and circularity. The concept of a fully integrated and decentralised Food–Water–Energy microgrid will from now on be referred to as a *Smarthood* (smart neighbourhood) and is depicted in Fig.

The benefit of implementing aquaponics into the Smarthoods concept is its potential to contribute to optimise integrated nutrient, energy and water flows (Fig.). This integration potential goes well beyond the already-mentioned

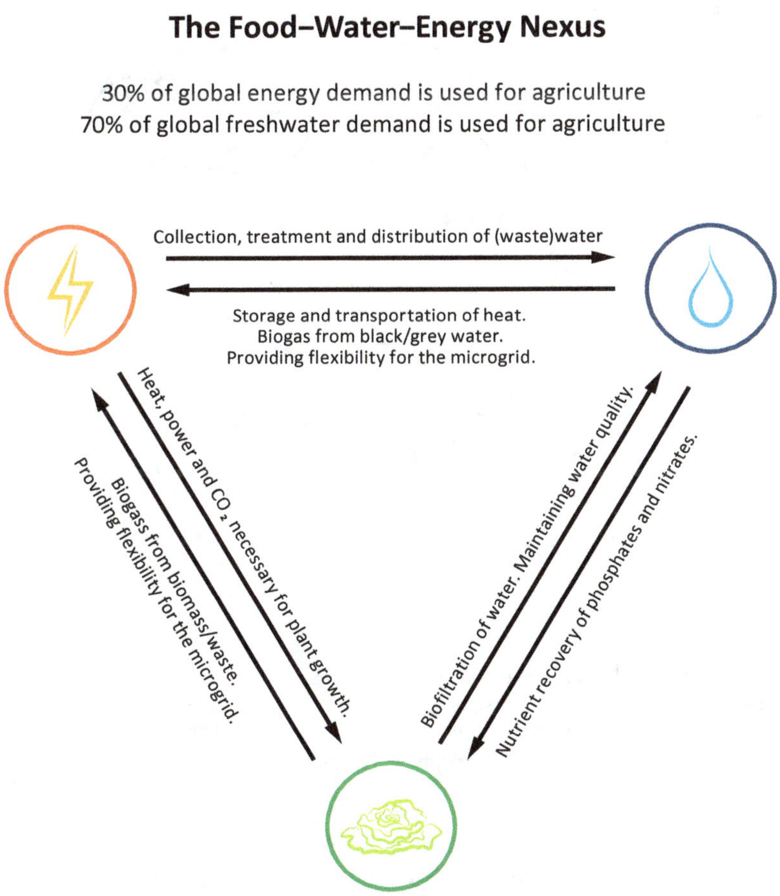

Fig. The Food–Water–Energy nexus shows the interplay between energy, water and food production (based on IRENA 2015)

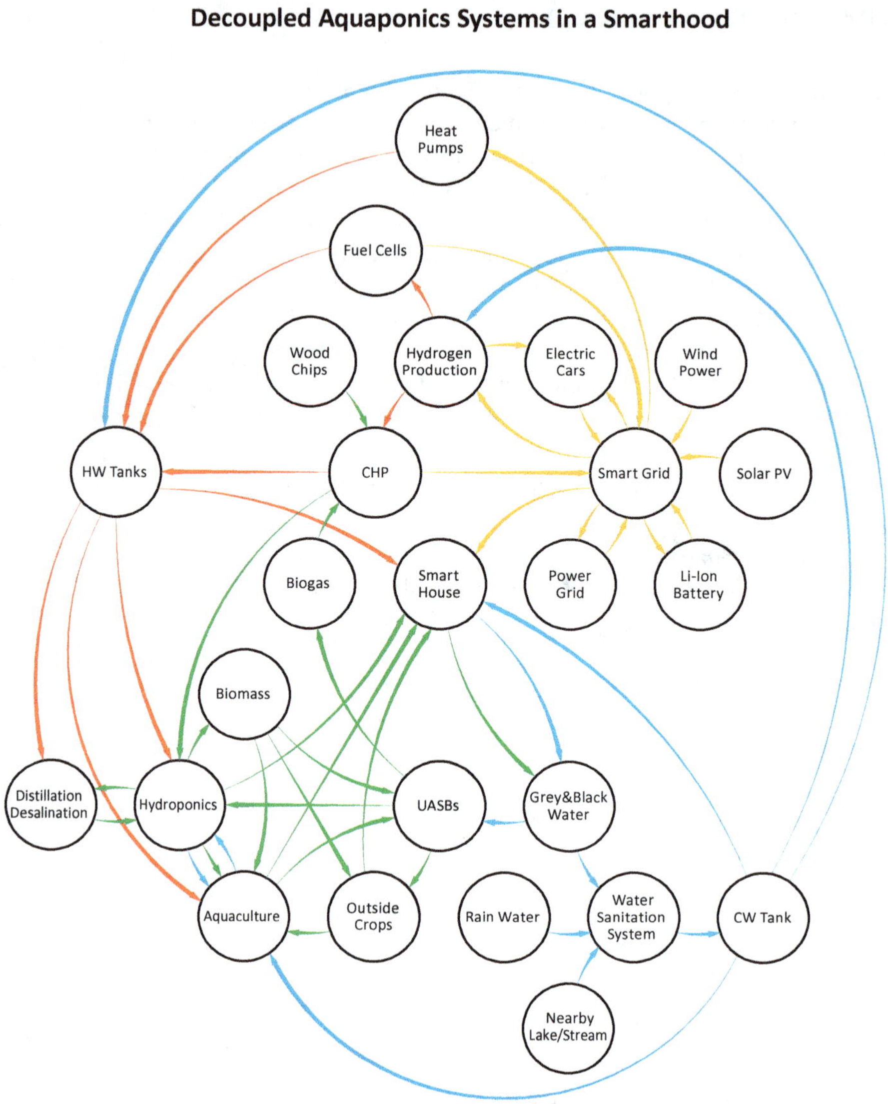

Fig. The integration of decoupled aquaponics systems (as described in Chapter) i n a decentralised local environment as designed for the Smarthoods concept. The green arrows show to what extent an aquaponics system can interact with the overall system. The red arrows represent heat flows, the blue arrows water flows and the yellow arrows power flows

crossovers between the energy and food systems. For instance, occurring biodegradable waste streams can be treated in anaerobic reactors (e.g. UASBs) and generate both biogas and bio-fertiliser (Goddek et al. 2018). Even the demineralized waste sludge can be utilised as liquid manure on conventional cropland.

Fig. Illustrated photo of *De Ceuvel* (source: Metabolic – www.metabolic.nl)

Example 1

An early example of an urban integrated aquaponic microgrid development is *De Ceuvel*, a previously abandoned shipyard in Amsterdam-North that has been converted into a self-sufficient office space and recreational hub. *De Ceuvel* serves as a testbed for new technologies and policies aimed at creating a circular economy. It features an all-electric microgrid including solar PV, heat pumps and peer-to-peer energy trading over the blockchain using their own energy token: the *Jouliette.*[1] A small aquaponic facility produces herbs and vegetables for the on-site restaurant. The same restaurant utilises biogas extracted from locally produced organic waste for their cooking activities as well as space heating. In addition, there is a lab present that is used for testing the water quality and extracting phosphates and nitrates.

Although *De Ceuvel* is currently not actively using the aquaponics facility to increase the flexibility of its microgrid, sensors are being installed to monitor the energy and nutrient flows in order to assess its performance. This data will be used to aid in the development of newer and smarter urban integrated aquaponics microgrids, such as the Smarthoods concept proposed in this chapter. Early use cases found in urban living labs like De Ceuvel are essential to the successful development of the Smarthoods concept (Fig.).

Although a holistic approach to urban FWE systems such as the *Smarthoods* concepts yields many benefits, the integration of aquaponics systems within microgrids remains very case-dependent. Aquaponic food production systems are characterised by a higher yield and a lower water, nutrient and energy footprint than conventional agricultural systems; however, they are also more costly to build. They are therefore best suited in locations that require high yields due to, for instance, space limitations. In dense urban areas, there may not always be sufficient space to build an aquaponics facility, whereas for rural areas the cost of land may be too low to warrant building a state-of-the-art aquaponics facility; a standard agricultural facility with lower financing costs and yield will be more suited in such cases. The most optimal use case for an integrated aquaponic facility is one where sufficient space is available, and a high yield per area is required to offset the cost of land use. Suburban neighbourhoods and other urban areas (e.g. an abandoned warehouse) are therefore most likely to see the first implementation of microgrids integrated with an aquaponic facility.

Goal

The goal of this research is to quantify the degree of self-sufficiency and flexibility for a microgrid integrated with a decoupled multi-loop aquaponics system.

Method

A neighbourhood of 50 households was assumed a 'Smarthood', with a decoupled multi-loop aquaponics facility present that is capable of providing fish and vegetables for all the 100 inhabitants of the Smarthood.

For the detailed modelling of the Smarthood, a hypothetical reference case of a suburban neighbourhood in Amsterdam was used, consisting of 50 households (houses) with an average household occupancy of 2 persons per household (100 persons total). In addition, one urban aquaponic facility consists of a greenhouse, aquaculture system, a UASB and a distillation unit. The dimensioning of the different components is motivated using data for a typical Dutch household and greenhouse (see Table).

The Energy System Model

An Energy System Model (ESM) was made that can simulate the energy flows of a wide range of components, whose main specifications are shown in Table. The ESM is capable of calculating energy flows for each component for each hour of the year.

Table Food and energy requirements per persona/household in the Netherlands

	Average (per capita/year)	Total (100 persons)	Source
Food			
Vegetable consumption (the Netherlands)	33 kg[a] (whereas 73 kg are recommended)	7300 kg	EFSA (2018)
Required greenhouse area	Approx. 4 m^2	400 m^2	Estimated based on min. consumption recommendation
Fish consumption	20 kg	2000 kg	FAO (2015)
Required aquaculture volume[b]	0.2 m^3	20 m^3	Estimated
Energy			
Household electricity consumption (Netherlands)	3000 kWh_e/house/year	150 MWh_e/year	CBS (2018)
Household heat consumption (Netherlands)	6500 kWh_{th}/house/year	325 MWh_{th}/year	CBS (2018)
RAS electricity consumption	0.05–0.15 kW_e/m^3	1–3 kW_e 8,76–28,26 MWh_e/year	(Espinal, pers. communication)

[a]The average Dutch person eats 50 kg of vegetables per year. However, only 33 kg of vegetables that can be grown in hydroponics systems, which are fruiting vegetables 31.87 g/day, brassica vegetables 22.11 g/day, leaf vegetables 12.57 g/day, legume vegetables 19.74 g/day, stem vegetables 4.29 g/day
[b]Considering a max. fish density of 80 kg/m^3

Table Production components

Component	Size	Specifications
Solar PV	40 $kW_{p,e}$	Eta: 0.15
Urban wind turbine	20 $kW_{p,e}$	Eta: 0.33
Heat pump	10 $kW_{p,e}$	COP: 4.0
CHP	20 $kW_{p,e}$	Eta_{el}: 0.24, $Eta_{th} = 0.61$
Fuel cell	10 $kW_{p,e}$	Eta: 0.55
Electrolyser	20 $kW_{p,e}$	Eta: 0.45
Battery	200 kWh	Eta: 0.90
Hot water tank	930 kWh	40–60°C
Hydrogen tank	1000 kWh	30 kg of H_2 storage

The energy system was modelled in MATLAB using energy profile data for Amsterdam obtained through DesignBuilder. The numerical time-series model incorporates a wide selection of energy technologies, listed in Table with their relevant specifications (Fig.).

The Energy System Model (ESM) uses simple conditional statements for the decision-making process, i.e. it is a rule-based control system. In the current version of this model, the control is centralised, with the objective of self-consumption

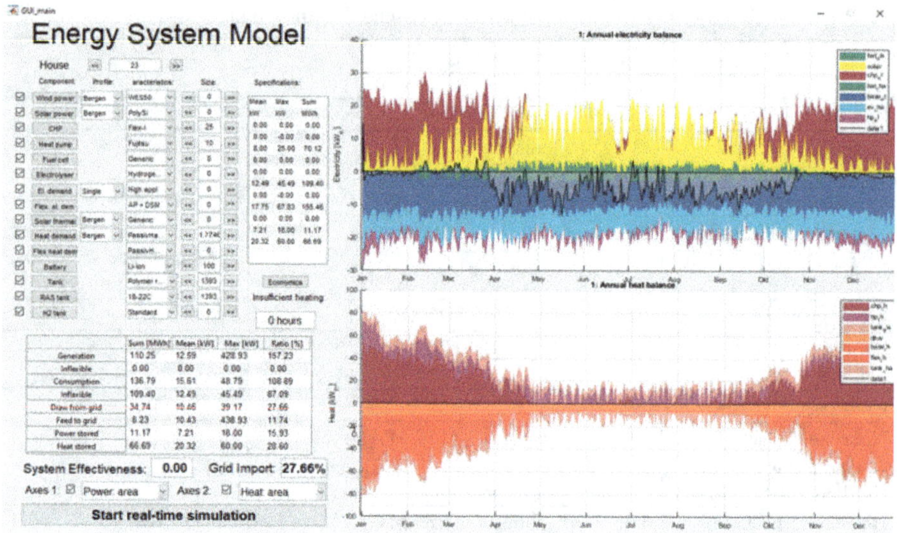

Fig. The aquaponics microgrid model (F. de Graaf 2018), showing the energy balances for power (upper diagram) and heat (lower diagram) for the reference case (Amsterdam)

maximisation for the system as a whole (in a future version, the control architecture will be decentralised, see Section). The conditional statements to achieve this can be stated as follows:

1. Keep the heat storage to a minimum.
2. Forecast the predicted inflexible electricity production and consumption.
3. (a) If the battery will be full, turn on flexible consumption.
   (b) If the battery will be empty, turn on flexible generation.

By keeping the heat storage to a minimum, the buffer for flexible energy balancing is maximised. If there is an overproduction of inflexible electricity (i.e. electricity production that cannot be flexibly scheduled or controlled, such as solar or wind), the heat pump can be turned on to create a buffer provided by hot water storage and the thermal mass of the aquaponic RAS system. Conversely, if there is an underproduction of electricity, flexible generation such as the CHP and the fuel cell can be turned on, thereby utilising the thermal storage capacity.

For both heat and power, the energy balance is equivalent to

$$P_{gen,flex} + P_{gen,inflex} + P_{grid} = P_{cons,inflex} + P_{cons,flex} + P_{storage} \tag{15.1}$$

Flexible generations include the heat pump, Combined Heat and Power (CHP) unit, fuel cell, battery and smart/flexible devices (e.g. aquaponic pumps). Wind, solar photovoltaics (PV) and solar collectors are classified as inflexible generation. Non-flexible devices make up the bulk of electricity consumption, especially in winter (due to the need for instant lighting) (Fig.).

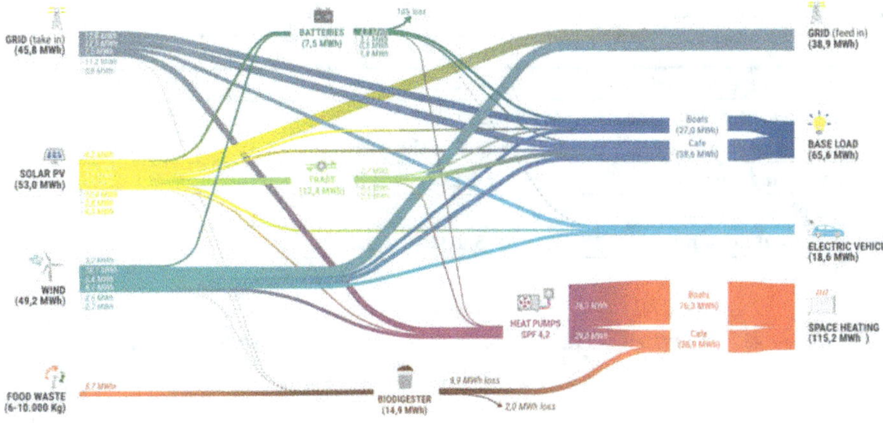

Fig. Example of the energy flows (Sankey diagram) of a possible integrated microgrid configuration at De Ceuvel (de Graaf 2018), including a biodigester for the production of biogas. This particular configuration does not include the Combined Heat and Power unit that is present within the Smarthood concept, nor does it take into account a large aquaponics facility

Results

The total electrical and thermal consumption of both the houses and the aquaponic greenhouse facility (modelled from the data in Tables) is shown in Table. The aquaponic greenhouse facility is responsible for 38.3% of power consumption and 51.4% of heat consumption. The power demand for an aquaponics facility integrated in a residential microgrid is therefore slightly over one-third of the total local energy demand, given that all of the residential energy and vegetable/fish production is done locally. The heat demand comprises roughly 50% of the total heat demand, which can be attributed for a large part to the distillation unit running on high-temperature water.

As can be seen in Figs., the Smarthoods energy system is capable of balancing production and demand most of the time. The total share of imported electricity from the grid is 4.62% for the reference case. At times, a slight imbalance of power can be observed, which can be attributed to suboptimal control for the current version of the model for the most part. The CHP, for instance, switches from an on- to off-state multiple times over the course of several hours, resulting in an overproduction of electricity. Such behaviour will not occur for a more optimised control system, since the CHP can be ramped down in coordination with the heat pump in order to deliver the precise amount of electricity and heat needed.

Flexibility

The system is highly flexible as a result of the CHP and the aquaponics facility with its flexible lighting and pumps, and high thermal buffering capacity, as well as the

Table Electrical and thermal load for different aspects of the microgrid

	Residential	Aquaponic facility
Electrical average demand	17.2 kW	10.2 kW
Electrical peak demand	47.6 kW_p	15.2 kW_p
Electrical total demand	**143.2 MWh/year**	**89.2 MWh/year**
Thermal average demand	37.1 kW	39.3 kW
Thermal peak demand	148.4 kW	121.2 kW
Thermal total demand	**325.0 MWh_{th}/year**	**344.2 MWh_{th}/year**

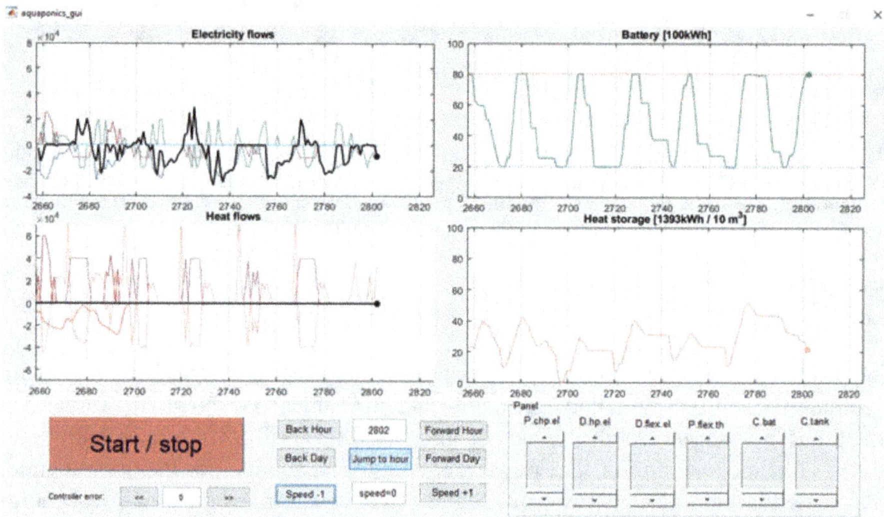

Fig. Time-series graphical diagrams for the power (top-left) and heat (bottom-left) energy balances (in W) of the Smarthood system. Storage capacity (in kWh) is indicated on the right side for power (top-right) and heat (bottom-right). The x-axis represents number of hours since the start of the year. The black line represents the imbalance of energy

battery, and the hydrogen system. The aquaponic system, especially, greatly increases the overall flexibility of the system, as it can function for a wide range of energy input, as can be derived from Table. As a result of this flexibility, the system manages to achieve near total (95.38%) power self-sufficiency and 100%heat self-sufficiency.

Discussion

Self-Sufficiency The energy system proposed for the Smarthood concept is capable of achieving near full grid-independence through the use of the flexibility provided by the various system components. The aquaponic system, especially, has a positive

Table Flexible demand of the aquaponic system

Component	Order of magnitude	Flexibility
Pumps	$0.05–0.15\ kW_e\ M^{-3}$ $1–3\ kW_e$ 8,76–28,26 MWh_e/year	Not all pumps have to run continuously. Main processes (oxygen control, ammonia control, CO_2 control, tank exchanges, suspended solids control) must run continuously. Smaller processes such as pH buffer dosing, backwash routines, water exchanges or back-up oxygenation do not have to run continuously
Lighting	80–150 W/m^2 With a capacity factor of 10–20% this leads to 28–105 MWh_e/year kW_e	Plants need ~4–6 h of darkness, the rest of the day they can be lit artificially. This leaves approx. 0 (summer) to 12 (winter) hours of flexible additional lighting
Space heating (underfloor) and aquaculture tank heating	444 kW_{th}/m^2/year 177,8 MWh_{th}/year	Due to the high thermal mass of the concrete floor and the large water volume in the RAS tank, the heat load is extremely flexible
Distillation unit	50 kW_{th} MWh_e/year 166,4 MWh_{th}/year	The distillation unit operates on hot water (70–90 °C) and can be operated with a significant degree of flexibility (MemSys 2017)

effect on the overall flexibility of the system. With 95.38% power self-sufficiency, this system performs better than any other economically feasible system assessed in previous research (de Graaf 2018).

Control Architecture Facilitating a decentralised local energy economy, such as the one proposed in the *Smarthoods* concept, requires a platform that keeps track of all the peer-to-peer transactions occurring within the neighbourhood. The corresponding peer-to-peer network can be classified as a multi-agent system (MAS) approach, in which multiple nodes (e.g. households or utility buildings) function as independent agents with their own objective (e.g. minimise cost or maximise energy saving) and corresponding decision-making process. Such a decentralised, multi-agent decision-making approach is necessary due to the complexity of the system. There is simply too much information and too many variables for the computation of a hierarchical, top-down and centralised control architecture.

Blockchain A blockchain-based multi-agent system control architecture could potentially provide the necessary framework to accommodate a decentralised peer-to-peer network. A vast number of distributed nodes ensure stability and security for the network, and an alternative to mining can be used: minting. With minting, tokens/coins are generated based on the data provided by a real-world device such as a smart energy metre. Provided that these sources of information can be trusted, i.e. that these devices can be tamper-proofed, a secure and independent ledger can be created in which various stakeholders can exchange goods (e.g. electricity) and services (e.g. demand-side management). Using smart contracts, complex services such as flexibility trading can be programmed into the control architecture of the system.

Internet of Things The constituent components within the Smarthood system, such as heat pumps, greenhouse lighting or the UASB, can all be controlled using Internet-connected sensors and actuators, known as Internet of Things. An IoT sensor network allows for the extensive acquisition of data, ranging from fish tank nutrient concentration to, for instance, battery load cycles, all on a per-time-step basis. This data can be used to verify the numerical model and optimise the dynamical control of the system.

Artificial Intelligence Optimising the control of the Smarthood system can be done by analysing the data using artificial intelligence algorithms, such as genetic programming (evolutionary algorithms) or machine reinforcement learning. With machine reinforcement learning, for instance, a set of actions and their influence on the environment are passed to the algorithm as input arguments, along with the current state of the system and a cumulative objective/cost function. An incrementally improving, heuristic decision-making process can be implemented at each household that will dynamically adapt to situations in order to find a near-optimal decision-making programme that will manage the energy flows within the house and the *Smarthood*. Each house can run such an algorithm, and as a result, a multi-nodal control system architecture, known as a multi-agent system (MAS), can be created that is relatively computationally inexpensive (compared to centralised control)—and close to optimal.

Legal Barriers The highly innovative nature of various aspects of the Smarthood concept, such as the polygeneration microgrid, the multi-loop aquaponic system and the unconventional urban planning requirements, brings along a unique set of challenges to overcome. For many of these challenges, the current regulatory framework is insufficient to accommodate the developments proposed in the Smarthoods concept.

Microgrids, for instance, work best when there is a local marketplace in which various *prosumers* (consumers that simultaneously produce power) can engage in frictionless peer-to-peer energy trading in a free market. Market forces will then work to create a local energy market in which a fluctuating energy price will result from local supply and demand. This price fluctuation will consequently incentivise smart energy solutions such as energy storage, demand-side management or flexible energy generation. In most EU countries, a free local market is currently impossible due to regulations; taxes have to be paid for every kWh that passes through the electricity metre, the electricity price for consumers is fixed and prosumers are not allowed to participate in the energy market without the intervention of a third party called the *aggregator*. With the expected increase in the development of microgrid projects, regulators will have to find ways to facilitate local energy markets in order to unlock the full potential of highly integrated microgrids (see Example .2).

Example 2
A recent advancement within the regulatory framework in the Netherlands is the introduction of the *experimenteerregeling,* an experimental law that allows a small number of carefully selected projects (such as de Ceuvel, example Z.1) to allow energy cooperatives to become their own distribution system operator, as if they were behind a single metre connection. This law is indicative of the awareness amongst Dutch regulatory bodies of previously mentioned legal barriers, and will therefore most likely lead to the current electricity law to be revised in the near future in order to better accommodate microgrid developments.

There are also some legal barriers in most EU countries with respect to reusing treated black water for fish and plant production, as it has to be ensured that human pathogens are fully eliminated. More information on the legal framework of aquaponics can be found in Chapter.

Conclusions

The goal of this research was to quantify the degree of flexibility and self-sufficiency that an aquaponics integrated microgrid can provide. In order to attain this answer, a neighbourhood of 50 households was assumed a 'Smarthood', with a decoupled multi-loop aquaponics facility present that is capable of providing fish and vegetables for all the 100 inhabitants of the Smarthood.

The results are promising: thanks to the high degree of flexibility inherent in the aquaponic system as a result of high thermal mass, flexible pumps and adaptive lighting, the overall degree of self-sufficiency is 95.38%, making it nearly completely self-sufficient and grid independent. With the aquaponics system being responsible for 38.3% of power consumption and 51.4% of heat consumption, the impact of the aquaponics facility on the total system's energy balance is very high.

Earlier research (de Graaf 2018) has indicated that it is very difficult to achieve self-consumption levels over 60% without relying on an external biomass source to drive a CHP. Even with this source included, the maximum techno-economically feasible self-consumption did not exceed 89%. In the Smarthood, biomass inputs for the CHP are partially derived from the aquaponic system itself, and the recycling of grey and black water. A higher self-consumption combined with a lower dependence on external biomass inputs, and a resulting self-consumption of 95%, makes the proposed aquaponic-integrated microgrid perform better from a self-sufficiency point of view than any other renewable microgrid known to the authors.

The authors of this chapter therefore strongly believe that with enough experimentation, integrating aquaponic greenhouse systems within microgrids yields great potential for creating highly self-sufficient Food–Water–Energy systems at a local level.

References

CBS (2018) Energieverbruik particuliere woningen; woningtype en regio's [WWW Document]. URL http://statline.cbs.nl/Statweb/publication/?VW=T&DM=SLNL&PA=81528NED&D1=a&D2=a&D3=0&D4=a&HD=161202-1617&HDR=T,G3&STB=G2,G1. Accessed 19 Oct 2018

de Graaf F (2018) New strategies for smart integrated decentralised energy systems

dos Santos MJPL (2016) Smart cities and urban areas—aquaponics as innovative urban agriculture. Urban For Urban Green 20:402–406. https://doi.org/10.1016/j.ufug.2016.10.004

EFSA (2018) The EFSA comprehensive European food consumption database | European [WWW Document]. URL https://www.efsa.europa.eu/en/food-consumption/comprehensive-database. Accessed 19 Oct 2018

FAO (2015) Statistical pocketbook - world food and agriculture

Goddek S, Körner O (2019) A fully integrated simulation model of multi-loop aquaponics: a case study for system sizing in different environments. Agric Syst 171:143–154

Goddek S, Delaide BPL, Joyce A, Wuertz S, Jijakli MH, Gross A, Eding EH, Bläser I, Reuter M, Keizer LCP, Morgenstern R, Körner O, Verreth J, Keesman KJ (2018) Nutrient mineralization and organic matter reduction performance of RAS-based sludge in sequential UASB-EGSB reactors. Aquac Eng 83:10–19. https://doi.org/10.1016/J.AQUAENG.2018.07.003

IRENA (2015) Renewable energy capacity statistics 2015, Masdar: IRENA

MemSys (2017) Membrane distillation system - MDS 500

Pudjianto D, Ramsay C, Strbac G (2007) Virtual power plant and system integration of distributed energy resources. IET Renew Power Gener 1:10. https://doi.org/10.1049/iet-rpg:20060023

Strbac G, Hatziargyriou N, Lopes JP, Moreira C, Dimeas A, Papadaskalopoulos D (2015) Microgrids: enhancing the resilience of the European megagrid. IEEE Power Energ Mag 13:35–43. https://doi.org/10.1109/MPE.2015.2397336

Yogev U, Barnes A, Gross A (2016) Nutrients and energy balance analysis for a conceptual model of a three loops off grid. Aquaponics Water 8:589. https://doi.org/10.3390/W8120589

9

Aquaponics and Social Enterprise

Sarah Milliken and Henk Stander

Abstract This chapter presents some examples of recent initiatives by social enterprises using aquaponics. Aquaponics offers an innovative form of therapeutic horticulture, which can provide employment and promote well-being for people with disabilities. If implemented as a program to be managed by local communities, aquaponic systems also have the potential to address issues such as food security and food sovereignty, especially in urban areas. Increasing public familiarity with aquaponics has seen a number of social ventures being set up around the world. However, the viability of these depends not only on stakeholder commitment, thorough market analysis, clear governance structures, and a robust business plan but also on external factors, such as the local political context and regulations.

Keywords Health · Well-being · Skills · Food security · Food sovereignty

Introduction

Social enterprises, as distinct from traditional private or corporate enterprise, aim to deliver products and services that cater to basic human needs. For a social enterprise, the primary motivation is not maximizing profit but building social capital; economic growth is therefore only part of a much broader mandate that includes social services such as rehabilitation, education and training, as well as environmental protection. There is growing interest in aquaponics among social enterprises, because it represents an effective tool to help them deliver their mandate. For example, aquaponics can integrate livelihood strategies to secure food and small

S. Milliken (✉)
School of Design, University of Greenwich, London, UK
e-mail: S.Milliken@greenwich.ac.uk

H. Stander
Department of Animal Sciences (Division of Aquaculture), University of Stellenbosch, Stellenbosch, South Africa
e-mail: hbs@sun.ac.za

incomes for landless and poor households. Domestic production of food, access to markets, and the acquisition of skills are invaluable tools for securing the empowerment and emancipation of women in developing countries, and aquaponics can provide the foundation for fair and sustainable socioeconomic growth (Somerville et al. 2014). This chapter presents some examples of recent initiatives by social enterprises using aquaponics.

Health, Well-being, and Skills

Aquaponics offers an innovative form of therapeutic horticulture, a nature-based approach that can promote well-being for people with mental health problems through using a range of green activities such as gardening and contact with animals. Over the past decade, a number of social enterprises have emerged that provide therapeutic horticulture programs for improving the well-being of local communities. The social enterprise approach builds on "social firms" by facilitating people with mental health problems to develop new skills and re-engage with the workplace. A social firm is a specific type of social enterprise where the social mission is to create employment, work experience, training, and volunteering opportunities, within a supportive and inclusive environment, for people who face significant barriers to employment and in particular for people with a disability (including mental ill health, and learning disability), abuse issues, a prison record, or homeless issues (Howarth et al. 2016).

There are particular qualities of the plant–person relationship that promote people's interaction with their environment and hence their health, functional level, and subjective well-being. Plants are seen to bestow non-discriminatory rewards on their carer without imposing the burden of an interpersonal relationship and, by responding to care or neglect, can immediately reinforce a sense of personal agency. The efficacy of practicing horticulture in a group context has also been demonstrated. Many people with mental and physical health problems face social exclusion because they do not have equal access to opportunities in society, including paid employment, housing, education, and leisure. Social networks such as those provided by community horticulture initiatives can act as buffers to stressors, provide a structure for acquiring skills, and validate and enhance an individual's self-worth (Diamant and Waterhouse 2010; Fieldhouse 2003).

To date there are few examples of social enterprises using aquaponics for therapeutic horticulture. In the United States, a small farming business called Green Bridge Growers in Indiana (www.greenbridgegrowers.org) is growing produce all year-round, primarily using aquaponics. The company now employs a number of individuals with autism spectrum disorder (ASD) and finds that the scheduling, precision, and monitoring required in aquaponics perfectly match with their skills (Fasciglione 2015). A core value of the business is to engage their workforce through leadership training, active participation, and team building, and provide them with the opportunity to gain new skills and competencies. Similarly, the ACRES Project (Adults Creating Residential and Employment Solutions; https://acresproject.org/aquaponics) in Pennsylvania uses aquaponics to provide horticultural therapy, employment, and community integration for adults with autism and intellectual disabilities. They are involved in all facets of the aquaponic system, from care and maintenance to harvest and sales, and the scheduled procedures and daily routines that aquaponics requires provide them with the stability and structure that they find reassuring. By fostering social, vocational, and self-advocacy skills, ACRES therefore uses aquaponics to help autistic individuals optimize their potential, develop practical life skills, increase social capacity, and transition to work and independence.

The FabLab Nerve Centre in Northern Ireland has set up a social enterprise aquaponic farm to teach people with learning difficulties entrepreneurial and digital skills. Using state-of-the-art digital equipment, such as 3D printers, CNC routers, and laser cutters, students will receive hands-on training and experience in a range of digital design and making techniques that will allow them to design, build, and operate an aquaponic farm. As part of the project, a newly created social enterprise will be developed by the young people, allowing them to sell the produce from the farm to local businesses, thereby developing their skills in social entrepreneurship, business, and marketing (www.nervecentre.org/news/fablab-nerve-centre-launches-aquaponic-digital-farm).

Solutions for Change, a social enterprise which is dedicated to solving family homelessness, runs Solutions Farms in California (www.solutionsfarm.org). The aquaponic farm provides training for homeless families in growing tilapia and seasonal leafy greens and herbs, which are then sold to local restaurants, markets, and schools. It functions as a laboratory for teaching important work values and preparing people for re-entry into the workplace, thereby raising hope, as well as produce.

Asociacíon Huerto Lazo (www.huertolazo.eu) is a social enterprise in the province of Malaga, Spain, which offers internships to young people from troubled backgrounds. The interns are given practical training in aquaponics in a safe environment. The catfish, tilapia, and tench are sold to El Sollo restaurant in Fuengirola (Fig.).

Fig. Aquaponic facilities at Asociacíon Huerto Lazo – anticlockwise from top left: catfish tanks in the aquaponic greenhouse; tilapia tanks with *Gynostemma pentaphyllum*, which is sold for medicinal purposes; the water filtration tanks at Huerto Lazo; Ulrich Eich demonstrating his aquaponic system (Photographs: Sarah Milliken)

Food Security and Food Sovereignty

Food security exists when all people, at all times, have physical and economic access to sufficient, safe, and nutritious food that meets their dietary needs and food preferences for an active and healthy life (Allison 2011). There are four food security pillars, which define, defend, and measure food security status locally, nationally, and internationally. These are food availability, food accessibility, food utilization, and food stability. Food availability is achieved when nutritious food is available at all times for people to access, while food accessibility is achieved when people at all times have the economic ability to obtain nutritious food available according to their dietary preferences. Food utilization is achieved when all food consumed is absorbed and utilized by the body to make a healthy active life possible, and food stability is achieved when all the other pillars are achieved (Faber et al. 2011).

Urban and peri-urban agriculture are increasingly recognized as a means by which cities can move away from current inequitable and resource-dependent food systems, reduce their ecological footprint, and increase their liveability (van Gorcum et al. 2019; Dubbeling et al. 2010). On account of being almost completely dependent on produce imported from other regions, urban consumers are particularly vulnerable to food insecurity. For those of low socioeconomic status, this dependence means that any fluctuation in food prices translates into limited purchasing power, increased food insecurity, and compromised dietary options. Community-based aquaponics enterprises offer a new model for blending local agency with scientific innovation to deliver food sovereignty and food security, by re-engaging and giving communities more control over their food production and distribution (Laidlaw and Magee 2016). If implemented as a program to be managed by local people, aquaponic systems have the potential to address food sovereignty. In turn, food security is boosted by consuming the fish, which is a significant source of protein, essential amino acids, and vitamins. Even when consumed in small quantities, fish can improve dietary quality by contributing essential amino acids, which are often missing or underrepresented in vegetable-based diets.

British social enterprise Byspokes Community Interest Company (CIC) set up a pilot aquaponic system and training program at the Al-Basma Centre in Beit Sahour, Occupied Palestinian Territories (OPT), a region where availability of space for food production is a serious problem, particularly in the urban areas and refugee camps. Even in agricultural areas, land access is being lost through Israeli controls and through effective annexation by the Israeli "Security Fence." Aquaponics therefore offers a water- and space-efficient solution to growing fresh, local produce, including a high-quality protein source (fish), thereby helping to combat malnutrition and food insecurity, while at the same time providing new opportunities for income generation. 40% of the population in the OPT (25% in the West Bank) are classed as "chronically food insecure", and unemployment stands at around 25%, with highs of 80% in some refugee camps. From an economic viewpoint, the project demonstrated that an aquaponic system could contribute significantly to household incomes and so help lift families out of poverty, while also providing a range of fresh vegetables and fish to families least able to afford such high-quality food (Viladomat and Jones 2011).

Since 2010, the Food and Agriculture Organization (FAO) of the United Nations has been implementing an *Emergency Food Production Support Project* for poor families in the Gaza Strip, where 11 years of Israeli sea, land, and air blockade, combined with low rainfall resulting in drought, have severely compromised the possibilities for domestic food production in one of the most densely populated areas of the world. With so many restrictions, fresh vegetables are expensive and hard to find. 97% of the Gaza Strip population are urban or camp dwellers and therefore do not have access to land. Poverty affects 53% of the population, and 39% of families headed by women are food insecure. Enabling families to produce their own affordable fresh food is therefore a highly appropriate and effective response to the current situation. Food-insecure female-headed households living in urban areas were given rooftop aquaponic units, and other units were installed in educational and community establishments. Having an aquaponic unit on their roof means that the women can simultaneously improve their household food security and income while still taking care of their children and homes. All of the beneficiaries have increased their household food consumption as a result (FAO 2016).

Through its Adaptive Agriculture Program, INMED Partnerships for Children is dedicated to establishing sustainable food programs that improve food security, conserve natural resources, promote strategies for adapting to climate change, and provide opportunities for income generation in developing countries. INMED has developed a simple and affordable aquaponic system for small-scale farmers, schools, government institutions, and home gardeners using easily accessible off-the-shelf local materials. Over the past decade, INMED has established a highly successful *Adaptive Aquaculture and Aquaponics Program* in South Africa, Jamaica, and Peru. In South Africa, INMED focuses on achieving food security and sustainable income generation by strengthening local capacity to understand and address climate change, while resolving interrelated issues of environmental degradation, increasing water scarcity, and poverty. It offers business-planning links to markets and assistance with applications for development grants and loans to expand enterprises. At the core of this far-reaching vision, in addition to intensive traditional cultivation, is aquaponics. Several projects have been successfully implemented in different provinces in the country. An aquaponic system was installed at the Thabelo Christian Association for the Disabled in a remote area of the Venda region in Limpopo province. Because INMED's system requires no heavy labour or complex mechanical systems, it is ideal for individuals with disabilities and those unable to perform traditional farming activities. Since the installation, the co-op has increased its revenue by more than 400%. Co-op members receive stable monthly salaries and have invested in breeding animals for additional revenue. Communities that have embraced this new way of farming have strengthened their ability to ensure food security and to provide new and adaptive opportunities for income generation (https://inmed.org) (Fig.).

Another good example of community upliftment in South Africa is Eden Aquaponics (www.edenaquaponics.co.za). Eden Aquaponics (Pty) Ltd. is the brainchild of Jack Probart who, with the realization that food security is fast becoming just as vital as a healthy economy, had the vision of developing a commercial business with a community focus. Using aquaponics to produce fish and vegetables in the Eden area of the Garden Route in the Western Cape, Eden Aquaponics supplies fish for consumption, as well as fingerlings for fish farming, and grows a variety of organic vegetables for distribution to the local farmers' markets, restaurants, and retailers. The Community Upliftment division manufactures and installs customized commercial systems of various sizes including DIY backyard aquaponic equipment, and supplies seedlings and fingerlings. They also teach less fortunate communities to become self-sufficient in growing, marketing, and selling their produce, thereby enabling previously unemployed people to develop skills, self-confidence, self-esteem, and the ability to provide for themselves.

Food Basket for Africa (www.foodbasketforafrica.org.za) is a similar initiative. They focus on men's development work, which is done through Men's Fraternity, by providing social and emotional counselling to men in rural areas. Men's Fraternity works in association with Effective Living Centres, a reputable counselling organization. Food Basket for Africa runs a number of agricultural development projects in rural areas around Southern Africa. Food tunnels (not hydroponic but mainly

Fig. Some illustrations of INMED's Community Projects in South Africa. (Photographs supplied by Janet Ogilvie from INMED)

wicking beds) are placed in communities, which adopt them, and initial training for tunnel care watering and fertilizing techniques is given. An aquaponic system was commissioned at one of the projects, in Kommetjie, Cape Town.

Issues of food security and food sovereignty are not only pertinent to the developing world. In Seville, Spain, social enterprise Asociacíon Verdes del Sur has set up an aquaponic greenhouse in the grounds of a school in Polígono Sur, the most socially deprived part of the city which is characterized by long-term unemployment and a high incidence of drug-related crime. The aquaponic unit is used as part of an environmental education program for local residents, including teaching the benefits of eating locally grown fresh food and developing skills for the unemployed (http://huertosverdesdelsur.blogspot.com). A prototype domestic unit has also been set up in the house of one of the local residents, Soledad (Fig.).

The Well Community Allotment Group (Crookes Community Farm) is a social enterprise run by volunteers in Sheffield, UK, that is on a mission to connect the local community with their food by actively involving them in its production, and by educating them about the benefits of local food. In 2018, the association was awarded an Aviva Community Fund Award in order to build an aquaponic unit which will be used to educate individuals, schools, youth groups, and other organizations (https://www.avivacommunityfund.co.uk/voting/project/pastwinnerprojectview/17-6291).

Fig. Aquaponic facilities in Polígono Sur – anticlockwise from top left: the aquaponic greenhouse at the school; Soledad with a frozen tilapia raised in her domestic unit; tomatoes and an aubergine saved for their seeds; the domestic aquaponic unit. (Photographs: Sarah Milliken)

In the United States, a number of social enterprises using aquaponic systems have been set up across the country as part of a growing social movement focusing on using urban agriculture to increase food security and community cohesion. One of the first was Growing Power, which was founded by Will Allen in 1995, with the objective of using urban agriculture as a vehicle for improving food security in central Milwaukee and for the long-term strengthening of its neighborhoods, and to give inner-city youngsters an opportunity to gain life skills by cultivating and marketing organic produce. Growing Power provided facilities or land, guidance in food growing, and overall project maintenance, and the produce was either donated to meal programs and emergency food providers or sold by the youngsters at local farm shops and farmers' markets, with the stipulation that one-quarter of the proceeds be returned to the local community (Kaufman and Bailkey 2000). By all accounts, Growing Power was doing exactly what they had set out to do: they were feeding, training, and exposing thousands of people to a more autonomous relationship with their food. But while their mission was being fulfilled, it carried significant costs. More money was exiting than entering Growing Power's doors, and by 2014, the social enterprise had a debt of more than $2 million (Satterfield 2018). Faced with insurmountable debt and legal pressure, Growing Power was eventually dissolved in 2017. However, the legacy of the enterprise lives on in the form of other social ventures that were inspired to start similar initiatives. One such venture which acknowledges Will Allen's influence is the Rid-All Green Partnership in Cleveland, Ohio, whose mission is to educate the next generation to not only learn to grow and eat fresh foods but also to operate and grow their own businesses in the food industry, ranging from selling fresh produce and fish to food distributors to processing and packaging fresh food products (https://www.greennghetto.org).

The urban agriculture movement in the United States has been fuelled by the US Department of Agriculture (USDA) Community Food Project (CFP) competitive grant program, which was established in 1996 with the aim of fighting food insecurity through the development of community food projects that promote the self-sufficiency of low-income communities. Since 1996, this program has awarded approximately $90 million in grants. One social enterprise which has benefited from this scheme is Planting Justice (www.plantingjustice.org) which built an aquaponic system on a vacant lot in East Oakland, California, which is run by former prison inmates. Twelve living wage jobs have been created, 5000 pounds (2268 kilos) of free produce has been given to the community, and the project has put $500,000 in wages and $200,000 in benefits back into the neighborhood (New Entry Sustainable Farming Project 2018).

The GrowHaus (https://www.thegrowhaus.org) was founded in 2009 as a social enterprise, which focuses on healthy, equitable, and resident-driven community food production; 97% of the food consumed in Colorado is produced out of state, and the neighborhood where The GrowHaus is located has been designated a "food desert" based on characteristics of low income, race/ethnicity, long distance to a grocery store, lack of access to fresh affordable food, and dependence on public transportation. The residents have come to rely on fast food, convenience stores, petrol stations, and food banks for the majority of their food staples. Due to these factors, many people face significant challenges in terms of food security and access, resulting in dramatic increases in related health issues. Initially in partnership with Colorado Aquaponics (www.coloradoaquaponics.com), and since 2016 independently, the GrowHaus operates a 3200 square foot (297 square meter) aquaponic farm, and the produce is sold through a weekly farm fresh food basket program at a price point comparable to Walmart, as well as to restaurants, with a portion donated to the local community. To help the transition to healthier eating, the GrowHaus also organizes free training and community events focused around food. In the 2016–2017 fiscal year, the GrowHaus generated an income of $1,204,070, of which $333,534 was earned income, and $870,536 was raised through government grants, charitable foundations, corporate contributions, and individual donations. With operating costs of $934,231, the net annual income was $269,839 (https://www.thegrowhaus.org/annual-report).

Trifecta Ecosystems (formerly Fresh Farm Aquaponics; http://trifectaecosystems.com) was founded in 2012 in Meriden, Connecticut. Their mission is to address urban food security by creating incentives for communities to grow their own food while also raising awareness about sustainable farming through education, workshops, and city projects. The enterprise employs six staff who provide aquaponic systems to organizations for educational purposes, workforce development, therapeutic gardening, and high-quality food production. The aquaponic systems range from commercial scale production facilities to small educational units for use in classrooms. In 2018, the South Central regional Water Authority (RWA) awarded a $500,000 grant to facilitate the creation of a series of custom-controlled environment agriculture aquaponic systems, an urban farming technology platform, and workforce training programs aimed at improving food security.

The SchoolGrown social enterprise (www.schoolgrown.org) was set up in 2014 by aquaponics enthusiasts who felt that children weren't getting enough hands-on experiences growing food and learning about their connection to the world about them. Situated next to the commercial aquaponics operation at Ouroboros Farms, California, the aquaponics 'classroom' is run by volunteers and is used to provide training. Their main focus, however, is on spreading aquaponic systems to schools and communities around the United States in order to teach sustainable agricultural practices, environmental stewardship, and resource conservation, and at the same time produce fresh and local food, thereby building a deeper connection between communities and the food they eat. The LEAF (Living Ecosystem Aquaponic Facility) is a 1800 square foot (167 square meter) greenhouse with a solar-powered aquaponic system that was specifically designed for this purpose. Costing $75,000, which includes salaries for two part-time staff responsible for maintaining the system and harvesting, the greenhouses are funded by a combination of a Community Supported Agriculture (CSA) vegetable box scheme, local community or business sponsorship, and crowdfunding. Each LEAF is intended to be financially self-sustaining through the generation of revenue from the produce.

The Viability of Aquaponics Social Enterprises

The examples above illustrate some of the different business models adopted by aquaponics social enterprises. Whether they will continue to thrive and grow or, like Growing Power, ultimately fail, remains to be seen. In the case of Growing Power, potential reasons for its collapse include Will Allen's inability to empower and retain an operational management team, and a lack of oversight by board members, which compromised the organization's financial health (Satterfield 2018). An in-depth analysis of two aquaponics social enterprises conducted in 2012–2013 revealed four distinct factors what were significant to their survival (Laidlaw and Magee 2016). Sweet Water Organics (SWO) began as an urban aquaponic farm in a large, disused, inner city industrial building in Milwaukee in 2008. It was funded primarily by its founders in order to develop creative capacity, employment opportunities, and chemical-free, fresh, and affordable food for the local community. In 2010, a new organization, Sweet Water Farms (SWF), was split from SWO, with the idea that they would grow as a mutually supportive, cohesive hybrid organization, including both a for-profit commercial urban farm (SWO) and a not-for-profit aquaponics "academy" (SWF). SWF managed volunteer operations and hosted training and education programs at the Sweet Water urban farm, while developing programs on a local (Milwaukee and Chicago), regional, national, and international scale. Sweet

Water had a loyal following among local restaurateurs and fresh food stores for its lettuce and sprouts produce, and sold its fish to a single wholesaler. However, the hybrid not-for-profit/for-profit enterprise model proved to be challenging, as both sides of the organization struggled to identify their role in relation to the other. While each side had a different structure relating to their operational character, and although their operations frequently overlapped, their strategic planning and visions sometimes did not. After 3 years of operation, SWO had still not managed to make a profit, and in 2011 the Milwaukee municipal government awarded a $250,000 loan on condition that 45 jobs would be created by 2014. In October 2012, SWO had 11–13 permanent employees, but was still being sustained through loans financing and equity investment. By June 2013, as loan repayments fell due and the job creation targets were not met, the for-profit arm of Sweet Water went into liquidation, and SWF took over as the primary operator of the Sweet Water urban farm. Currently, SWF operates entirely as an educational and advisory enterprise run by volunteers and a small team of part-time employees, and no longer supplies restaurants with produce (Laidlaw and Magee 2016).

The Centre for Education and Research (CERES) in Melbourne, Australia, opened its aquaponics facility in 2010. The system was designed as a suboptimized commercial system with the production capacity to support a single wage for the farmer who maintains it. This wage varies based on how much he/she produces, with the vegetables being sold through the CERES Fair Food organic box delivery service. The scale of the operation does not generate a return that would permit the setting up of a fish-processing facility. Stakeholders at Sweet Water Farms and CERES identified that the principal factor behind their survival was ongoing commitment, in the form of continued support of personnel with technical and business management skills combined with an enduring leadership, and the willingness of the stakeholders to remain involved and prepared to cooperate without strong financial incentives. The second factor was the local political context. While the city of Milwaukee supported Sweet Water both through policy initiatives and direct financial aid, which allowed it to expand its fixed assets and human resources, build market awareness, and acquire a sizeable regular commercial customer base, the CERES project had little such support, beyond an initial grant, and it had struggled to generate revenue, which would have allowed it to expand. Costs of compliance and licensing also made it difficult to engage with local markets in more than a token way, which dampened its motivation to market and sell the produce, and made it untenable for the operation to develop beyond a small part-time income-generating enterprise. The third factor was the availability of markets for urban aquaponics produce. While urban aquaponics is attractive to a customer base that is increasingly responsive to issues of food security and ethical consumption, such as in Milwaukee, this was not the case in Melbourne. The final factor was diversification. Both CERES and SWO/SWF benefitted from translating social and technical experimentation into a range of training and educational services. SWO/SWF, being a larger concern, obviously had greater capacity for developing these services, and these proved vital in sustaining the social enterprise when commercial plans failed to materialize (Laidlaw and Magee 2016).

Conclusions

In "Ten technologies which could change our lives" (European Parliamentary Research Service, 2015), aquaponic systems were singled out as a solution for developing innovative and sustainable food sources for Europe which, through shortening of supply chains, could improve food security and food systems resilience. However, the technology is still newly emerging and as yet relatively undeveloped, and as the study by Laidlaw and Magee (2016) highlights, the viability of an aquaponics social enterprise depends not only on stakeholder commitment, thorough market analysis, clear governance structures, and a robust business plan, but also on external factors, such as the local political context and regulations.

References

Allison EH (2011) Aquaculture, fisheries, poverty and food security. Security 61

Diamant E, Waterhouse A (2010) Gardening and belonging: reflections on how social and therapeutic horticulture may facilitate health, wellbeing and inclusion. Br J Occup Ther 73(2):84–88. https://doi.org/10.4276/030802210X12658062793924

Drangert JO (1998) Fighting the urine blindness to provide more sanitation options. Water SA 24:157–164

Dubbeling M, de Zeeuw H, van Veenhuizen R (2010) Cities, poverty and food: multi- stakeholder policy and planning in urban agriculture. Practical Action Publishing Ltd., Colchester

European Parliamentary Research Service (2015) Ten technologies which could change our lives: Potential impacts and policy implications. http://www.europarl.europa.eu/EPRS/EPRS_IDAN_527417_ten_trends_to_change_your_life.pdf

Faber M, Witten C, Drimie S (2011) Community-based agricultural interventions in the context of food and nutrition security in South Africa. S Afr J Clin Nutr 24:21–30

FAO (2016) Urban agriculture in the Gaza Strip through vertical gardens and aquaponics. Food and Agriculture Organization of the Unuted Nations, Rome. http://www.fao.org/fileadmin/templates/FCIT/PDF/Fact_sheet_on_aquaponics_Final.pdf

Fasciglione M (2015) Corporate social responsibility and the right to employment of persons with disabilities. In: Fina VD, Cera R (eds) Protecting the Rights of People with Autism in the Fields of Education and Employment. Springer, Cham, pp 171–193. https://doi.org/10.1007/978-3-319-13791-9_7

Fieldhouse J (2003) The impact of an allotment group on mental health clients' health, wellbeing and social networking. Br J Occup Ther 66(7):286–296

Howarth ML, McQuarrie C, Withnell N, Smith E (2016) The influence of therapeutic horticulture on social integration. J Public Ment Health 15(3):136–140. https://doi.org/10.1108/JPMH-12-2015-0050

Kaufman J, Bailkey M (2000) Farming inside cities: entrepreneurial urban agriculture in the United States. Lincoln Institute of Land Policy working paper. http://www.urbantilth.org/wp-content/uploads/2008/10/farminginsidecities.pdf

Laidlaw J, Magee L (2016) Towards urban food sovereignty: the trials and tribulations of community-based aquaponics enterprises in Milwaukee and Melbourne. Local Environ 21 (5):573–590. https://doi.org/10.1080/13549839.2014.986716

New Entry Sustainable Farming Project (2018) Community food projects: indicators of success FY 2017. https://nesfp.org/sites/default/files/uploads/ios_2017_final_draft_april_13_2018.pdf

Satterfield S (2018) Behind the rise and fall of Growing Power. Public Radio International, March 2018. https://www.pri.org/stories/2018-03-16/behind-rise-and-fall-growing-power

Somerville C, Cohen M, Pantanella E, Stankus A, Lovatelli A (2014) Small-scale aquaponic food production. Integrated fish and plant farming. FAO fisheries and aquaculture technical paper no. 589. FAO, Rome

van Gorcum et al (2019) https://link.springer.com/article/10.1007/s10499-019-00379-1

Viladomat L, Jones P (2011) Development of aquaponic systems for space and water efficient food production in the Occupied Palestinian Territories – Pilot project final report. http://www.byspokes.org/wp-content/uploads/documents/basma_report.pdf

10

Aquaponics in the Built Environment

Gundula Proksch, Alex Ianchenko, and Benz Kotzen

Abstract Aquaponics' potential to transform urban food production has been documented in a rapid increase of academic research and public interest in the field. To translate this publicity into real-world impact, the creation of commercial farms and their relationship to the urban environment have to be further examined. This research has to bridge the gap between existing literature on growing system performance and urban metabolic flows by considering the built form of aquaponic farms. To assess the potential for urban integration of aquaponics, existing case studies are classified by the typology of their building enclosure, with the two main categories being greenhouses and indoor environments. This classification allows for some assumptions about the farms' performance in their context, but a more in-depth life cycle assessment (LCA) is necessary to evaluate different configurations. The LCA approach is presented as a way to inventory design criteria and respective strategies which can influence the environmental impact of aquaponic systems in the context of urban built environments.

Keywords Aquaponics classification · Urban aquaponics · Enclosure typologies · Greenhouses · Indoor growing · Controlled environment agriculture · Life cycle assessment

Introduction

Aquaponics has been recognized as one of "ten technologies which could change our lives" by merit of its potential to revolutionize how we feed growing urban

G. Proksch (✉) · A. Ianchenko
Department of Architecture, College of Built Environments, University of Washington, Seattle, WA, USA
e-mail: prokschg@uw.edu

B. Kotzen
School of Design, University of Greenwich, London, UK
e-mail: b.kotzen@greenwich.ac.uk

populations (Van Woensel et al. 2015). This soilless recirculating growing system has stimulated increasing academic research over the last few years and inspired interest in members of the public as documented by a high ratio of Google to Google Scholar search results in 2016 (Junge et al. 2017). For a long time, aquaponics has been primarily practiced as a backyard hobby. It is now increasingly used commercially due to strong consumer interest in organic, sustainable farming methods. A survey conducted by the CITYFOOD team at the University of Washington in July 2018 shows that the number of commercial aquaponic operations has rapidly increased over the last 6 years. This focused search for aquaponic operations identified 142 active for-profit aquaponic operations in North America. Based on online information, 94% of the farms have started their commercial-scale operation since 2012; only nine commercial aquaponic farms have been in operation for more than 6 years (Fig.).

Most of the surveyed aquaponic operations are located in rural areas and are often connected to existing farms to take advantage of low land prices, available infrastructure, and conducive building codes for agricultural structures. Regardless, a growing number of aquaponic operations are also located in cities. Due to their

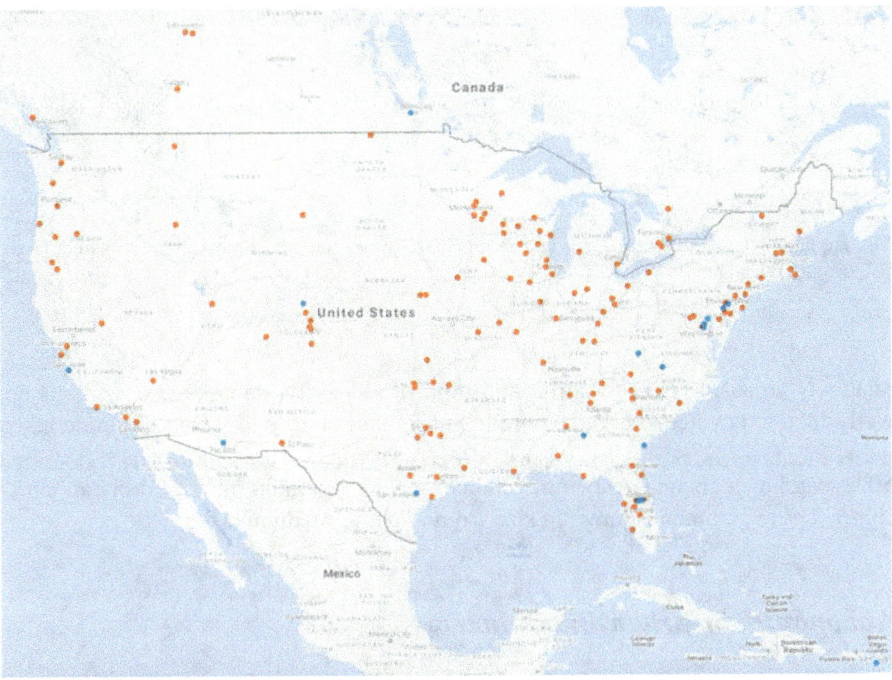

Fig. Existing aquaponic practitioners in North America, 142 commercial companies (red) and 17 research centers (blue), (CITYFOOD, July 2018)

Fig. Aquaponics across Europe: 50 research centers (blue) and 45 commercial companies (red). (EU Aquaponics Hub 2017)

relatively small physical footprint and high productivity, aquaponic operations are well suited to practice in urban environments (Junge et al. 2017). Surveys undertaken under the auspices of the European Union (EU) Aquaponics Hub in 2017 identified 50 research centers and 45 commercial companies operating in the European Union (Fig.). These companies range in size from small to medium sized.

Aquaponics in Urban Environments

Space is a valuable commodity in cities. Urban farms have to be resourceful to find available sites such as vacant lots, existing rooftops, and underutilized warehouses that are affordable for an agricultural business (de Graaf 2012; De La Salle and Holland 2010). Urban aquaponic farms need to balance higher production costs with competitive marketing and distribution advantages that urban locations offer. The largest benefit for locating aquaponic operations in cities is a growing consumer market with an interest in fresh, high-quality and locally grown produce. When complying with local regulations for organic produce, urban farms can achieve premium prices for their aquaponically grown leafy greens, herbs, and tomatoes (Quagrainie et al. 2018). Unlike hydroponics, aquaponics also has the capacity to produce fish, further enhancing economic viability in an urban setting which often has diverse dietary needs (König et al. 2016). Urban aquaponic farms can also save some operational costs by reducing transportation distance to the consumer and reducing the need for crop storage (dos Santos 2016).

Urban environmental conditions can also be advantageous for aquaponic farms. Average temperatures in cities are higher than in rural surroundings (Stewart and Oke 2010). In colder regions particularly, farms can benefit from a warmer urban climate, which can help reduce heating demand and operational costs (Proksch 2017). Aquaponic farms that are integrated with the building systems of a host building can further utilize urban resources such as waste heat and CO_2 in exhaust air to benefit the growth of plants as an alternative to conventional CO_2 fertilization. Urban farms can also help mitigate the negative aspects of the urban heat island effect during the summer months. The additional vegetation, even if grown in greenhouses, helps to reduce the ambient temperature through increased evapotranspiration (Pearson et al. 2010). In aquaponics, the use of recirculating water infrastructure reduces overall water consumption for the production of both fish and lettuce and can, therefore, have a positive effect on the urban water cycle. Aquaponically-grown produce strives to close the nutrient cycle, thereby avoiding the production of agricultural run-off. Through smart resource management within major environmental systems, aquaponics helps to reduce excessive water consumption and eutrophication usually created by industrial agriculture.

Aquaponics as Controlled Environment Agriculture (CEA)

Traditional agricultural techniques to extend the natural crop-growing season range from minimal environmental modifications, such as temporary hoop houses used on soil-based fields, to full environmental control in permanent facilities that allow for year-round production regardless of the local climate (Controlled Environment Agriculture 1973). The latter strategy is also known as controlled environment agriculture (CEA) and includes both greenhouses and indoor growing facilities. In addition to controlling the indoor climate, CEA also significantly reduces the risk of crop loss to natural calamities and the need for herbicides and pesticides (Benke and Tomkins 2017). Most aquaponic operations are conceived as CEA since they combine two complex growing systems (aquaculture and hydroponics), which both require controlled growing conditions to guarantee optimal productivity. Additionally, CEA enables year-round production to amortize high investment in aquaponic infrastructure and achieve premium crop prices at the market outside of the natural growing season. The performance of aquaponic farm enclosures is highly dependent on local climate and seasonal swings (Graamans et al. 2018).

As aquaponics is a relatively young discipline, most of the existing research is focused at the system level – for example, studies evaluating the technical integration of aquaculture with hydroponics in different configurations (Fang et al. 2017; Lastiri et al. 2018; Monsees et al. 2017). Whilst individual aquaponic system components and their interactions can still be further optimized for productivity, their performance within a controlled environment envelope has not been comprehensively addressed. Recent research in CEA has begun to assess hydroponic system performance in tandem with built environment performance, although there is only one study to date that models aquaponic system performance in a controlled envelope (Benis et al. 2017a; Körner et al. 2017; Molin and Martin 2018a; Sanjuan-Delmás et al. 2018).

Aquaponics Research Collaborations

The current expansion in interest in aquaponics led to the creation of several interdisciplinary aquaponics related research collaborations funded by the European Union (EU). The COST FA1305 project, which created the EU Aquaponics Hub (2014–2018) brought together aquaponics research and commercial producers to better understand the state of the art in aquaponics and to generate coordinated research and education efforts across the EU and around the world. Innovative Aquaponics for Professional Application (INAPRO) (2014–2017), a consortium of 17 international partners, aimed to advance current approaches to rural and urban aquaponics through the development of models and construction of prototypical greenhouses. The project CITYFOOD (2018–2021) within the Sustainable Urban Growth Initiative (SUGI), co-funded by the EU, Belmont Forum, and respective science foundations, investigates the integration of aquaponics in the urban context and its potential impact on global challenges of the food-water-energy nexus.

Classification of Controlled Environment Aquaponics

The term aquaponics is used to describe a wide range of different systems and operations, greatly varying in size, technology level, enclosure type, main purpose, and geographic context (Junge et al. 2017). The first version of the classification criteria for aquaponic farms included stakeholder objectives, tank volume, and parameters describing aquaculture and hydroponic system components (Maucieri

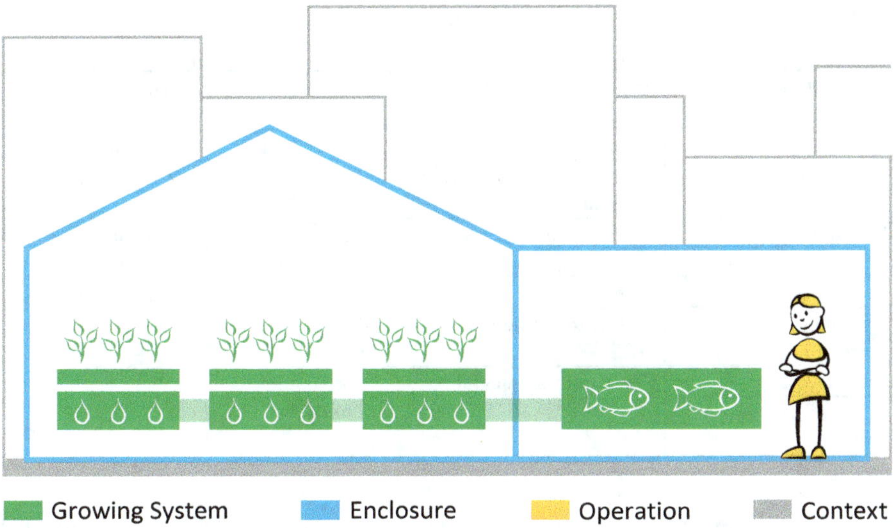

Fig. Classification criteria for identifying aquaponic farm types

et al. 2018). Additional work was undertaken by a large group of researchers to further define aquaponics and to present a nomenclature based on international consensus (Palm et al. 2018). This led to a comprehensive discussion on system types and scales and most importantly a definition of aquaponics which is: "*the majority (> 50%) of nutrients sustaining the optimal plant growth must derive from waste originating from feeding the aquatic organisms.*" However, both definitions focus on the growing systems and do not consider other essential aspects of a functioning commercial aquaponics farm. As aquaponic operations become part of local economies, classification criteria identified by interdisciplinary research in fields like architecture, economics, and sociology will also become essential.

This classification proposal focuses on the emerging field of commercial aquaponic operations through the lens of the built environment. The key characteristics that describe an aquaponic operation fall into four different categories: growing system, enclosure, operation, and context (Fig.). These categories for classifi-cation criteria impact one other across scales, where growing system configurations can affect the contextual performance of the farm as a business, or local market demands can determine the type of crop grown in the aquaponic system. Some farm classification criteria are relevant on all scales, such as "size," measured in tank volume, growing area, number of employees, and annual revenue (Table).

- *Growing system* classification criteria describe the configuration of the interconnected aquaculture and hydroponic system. This includes specifications for the physical components that enable water and nutrient recirculation (such as water tanks, filters, pumps, and piping), living organisms that transform available nutrients at different stages (including fish, plant, and microorganism species) and

Table Possible classification criteria for aquaponic farm types

Growing system	Enclosure	Operation	Context
Aquaculture system type	Enclosure typology	Purpose	Geographical location
Fish species	Structural system	Stakeholders	Physical context
Water temperature	Envelope assembly cover material	Business model	Environmental impact
Filtration system	Heating/cooling systems	Labor distribution	Socioeconomic context
Feed type	Light source	Funding type	Social impact
Hydroponic system type	Ventilation system	Marketing scheme	
Crop species	Host building integration	Distribution model	
Water distribution system			

values describing the physical performance of the system, such as temperature, pH levels, oxygen/carbon content, and electrical conductivity (Alsanius et al. 2017).

- *Enclosure* classification criteria define characteristics of the buildings that house the growing systems, at the next scale. Most aquaponic farms use CEA enclosures that vary by identifying typology (such as a greenhouse or warehouse), structural system, heating and cooling systems, lighting, ventilation, and humidity control systems (Benke and Tomkins 2017).
- *Operations* classification criteria describe how each aquaponic farm operates as a business and farm, which includes human expertise and labor input necessary for growing and selling produce. Criteria in this section include funding type, business structure and management, labor requirements and division, marketing scheme, produce distribution model, and overall purpose of the aquaponics facility.
- *Context* classification criteria, at the largest scale, describe the geographic location, physical context, urban integration, and overall social impact of aquaponic farms. Context criteria describe how an aquaponic farm is part of the urban food chain and built environment, capable of influencing economic growth, social involvement and large-scale environmental impacts on a city-wide scale (dos Santos 2016).

Enclosure Typologies and Case Studies of Commercial Farms

This further investigation focuses on defining aquaponic classification criteria at the enclosure level to complement existing system-level definitions. The enclosure types discussed here work with different construction systems, levels of technological control, passive climate control strategies, and energy sources to achieve an appropriate indoor climate. The best application of each enclosure typology depends primarily on the size of operation, geographic location, local climate, targeted fish and crop species, required parameters of the systems it houses, and the budget. This study identifies five different enclosure typologies and defines the characteristics of indoor spaces that house aquaculture infrastructure.

Greenhouse Typologies

This classification includes four categories of greenhouses – medium-tech greenhouses, passive solar greenhouses, high-tech greenhouses, and rooftop greenhouses – that are applicable to commercial-level aquaponic operations (Table). Existing greenhouses may not exactly fit a single typology, but fall within a spectrum from medium-tech to high-tech by selectively incorporating active and passive environmental control techniques.

Medium-Tech Greenhouses Greenhouses with intermediate levels of technology to control the indoor climate include freestanding or gutter-connected Quonset (Nissen hut type), hoop house (polytunnel) and even-span greenhouses. They are usually covered with double polyethylene film (PE) or rigid plastic panels, such as acrylic panels (PMMA) and polycarbonate panels (PC). These greenhouses are less expensive to install, though film cladding needs to be replaced frequently due to rapid deterioration caused by constant exposure to UV radiation (Proksch 2017). These greenhouses protect crops from extreme weather events and to some extent pathogens, but they offer only a limited level of active climate controls. Instead, they rely on solar radiation, simple shading systems, and natural ventilation. With their limited ability to modify growing conditions within a certain range, medium-tech greenhouses are rarely used for housing aquaponic farms in cold climates. This is because the high initial investment into the hydroponic and aquaculture components requires a stable environment and reliable year-round production to be commercially viable.

Aquaponic operations in warmer climates have successfully demonstrated the use of medium-tech greenhouses that employ evaporative cooling and simple heating systems. For example, Sustainable Harvesters in Hockley, Texas, USA uses a simple Quonset greenhouse (12,000 sf/1110 m^2) for year-round lettuce production without relying on extensive supplemental heating or lighting. Ouroboros Farms in Half Moon Bay, California, USA uses an existing greenhouse (20,000 sf/1860 m^2) t o produce lettuce, leafy greens, and herbs (Fig.). Due to the mild climate, the farm uses primarily static shading and little supplemental heating and cooling. Both farms, as many smaller medium-tech operations, place their fish tanks in the same greenhouse space as the hydroponic crop growing system. The farms grow fish species that tolerate a wide temperature range (tilapia) and shade aquaculture tanks to prevent overheating and algae growth.

Table Comparison of case studies by enclosure typologies

CEA type	Case studies	Construction system	Controls	Growing season[a] and latitude	Hardiness zone[b]
Medium-tech greenhouses	Ouroboros Farms, Half Moon Bay, CA, USA (20,000 sf/1860 m²)	Existing, gutter-connected GH with two even spans, clad with single-pane glass, fish tanks in GH	Static shading, shading curtains	319 days/ 10.6 months	10a 30 to 35 °F
				37.5° N	−1.1 to 1 .7 °C
	Sustainable Harvesters, Hockley, TX, USA (12,000 sf/1110 m²)	Quonset frame, multi-tunnel (3) GH, clad with PE- film and rigid plastic panels, fish tanks in GH	Evaporative cooling, forced ventilation	272 days/ nine months	8b 15 to 20 °F
				30.0° N	−9.4 to 6 .7 °C
Passive solar greenhouses	Aquaponic solar green-house, Neuenburg am Rhein, Germany (2000 sf/180 m²)	(Chinese) solar greenhouse, with adobe wall as additional thermal mass, clad with ETFE film, fish tanks in GH	Custom-built photovoltaic modules for shading and energy production	202 months/ 6.6 months	8a 10–15 °F
				47.8° N	−12.2 to −9 .4°C
	Eco-ark greenhouse at Finn & Roots, Bakersfield, VT, USA (6000sf/ 560 m²)	Solar green-house, earth sheltered, steep angle of south facing roof (ca. 60°), thick insulation, spe-cial solar collecting glaz-ing, fish tanks in northern, sub-terranean side	Wood-fuelled radiant heat, energy curtain, ventilation with stack-effect, supple-mental LED lighting	108 days/ 3.6 months	4a −30 to −25 °F
				44.8° N	−34.4 to −31.7 °C
High-tech greenhouses	Superior Fresh Farms, Hixton, WI, USA (123,000 sf/11,430 m²)	Venlo-style, gutter-connected, (20 × 41 bays), clad with glass, fish tanks in separate building	Computer-con-trolled CEA environment, supplemental LED lighting,	122 days/ 4.1 months	4b −25 to −20 °F
				44.4° N	−31.7 to −28.9 °C
	Blue Smart Farms, Cobbitty, NSW, Australia (53,800 sf/5000 m²)	Venlo-style, gutter-connected, (14 × 18 bays), clad with glass, two-story con-struction, fish tanks on the lower level	Computer-con-trolled CEA environment biological pest control	300 days/ 10 months	9b 25 to 30 °F
				34.0°S	−3.9 to −1.1 °C

(continued)

Table (continued)

CEA type	Case studies	Construction system	Controls	Growing season[a] and latitude	Hardiness zone[b]
Rooftop greenhouses	Ecco-jäger Aquaponik Dachfarm, Bad Ragaz, Switzerland (12,900 sf/1200 m^2)	Venlo-style, gutter-connected, (7 × 13 bays), clad with glass, fish tanks on the lower level	CEA environment, supplemental LED lighting, use of exhaust heat from cooling facility	199 days/ 6.6 months	7b 5 to 10 °F
				47.0° N	−15.0 to −12.2 °C
	BIGH's Ferme abattoir, Brussels, Belgium (21,600 sf/2000 m^2)	Venlo-style, gutter-connected, (15 × 10 bays), clad with glass, fish tanks on the lower level	CEA environment, supplemental LED lighting	224 days/ 7.3 months	8b 15 to 20 °F
				50.8° N	−9.4 to −6 .7 °C
Indoor growing spaces	Urban Organics, Schmidt's Brewery, St. Paul, MN, USA (87,000 sf/8080 m^2)	Steel-frame warehouse, highly insulated, stacked growing, fish tanks in separate space	Fluorescent UV lighting, computer-controlled CEA environment	140 days/ 4.7 months	4b −25 to −20 °F
				45.0° N	−31.7 to 28.9 °C
	Nutraponics, Sherwood Park, AB, Canada (10,800 sf/1000 m^2)	Steel-frame warehouse, highly insulated, stacked growing, fish tanks in separate space	LED lighting, computer-controlled CEA environment	121 days/ 4 months	4a −30 to −25 °F
				53.5° N	−34.4 to −31.7 °C

[a]Frost-free growing season, National Gardening Association, Tools and Apps, https://garden.org/apps/calendar/

[b]Based on the USDA Hardiness Zone Map, which identifies the average annual minimum winter temperature (1976–2005), divided into 10° F zones. Plant Maps, https://www.plantmaps.com/index.php

Passive Solar Greenhouses This greenhouse type is designed to be solely heated by solar energy. Substantial thermal mass elements, such as a solid north-facing wall, store solar energy in form of heat that is then re-radiated during colder periods at night. This approach buffers air temperature swings and can reduce or eliminate the need for fossil fuels. Solar greenhouses have a transparent south-facing side and an opaque, massive, highly insulated north-facing side. The integration of large volumes of water in form of fish tanks is an asset for the thermal performance of this greenhouse type. Furthermore, the tanks can be located in areas of the greenhouse that are less suited for plant cultivation or partly submerged into the ground for added thermal stability.

Fig. Ouroboros Farms (Half Moon Bay, California, USA)

The Aquaponic solar greenhouse (2000 sf/180 m^2), developed and tested by Franz Schreier, has proven as a suitable environment for housing a small aquaponic system in southern Germany. The greenhouse collects solar energy through its south-facing arched roof and wall clad with ethylene tetrafluoroethylene (ETFE) film. Heat is stored in partially submerged fish tanks, floor, and adobe-clad northern wall to be dissipated at night. The greenhouse's custom-built photovoltaic (PV) panels transform solar radiation into power. Located in the colder climate of Vermont, USA, the Eco-Ark Greenhouse at the Finn & Roots farm (6000 sf/560 m^2) houses an aquaponic system that works with a similar passive solar approach. The greenhouse has a steep (approx. 60°) south-facing transparent roof with special solar-collecting glazing (Fig.). Its highly insulated, opaque northern side is sub-merged into a hillside and houses the fish tanks. In addition to these passive controls, the Eco-Ark has a radiant floor heating that supplements heating during the coldest seasons.

High-Tech Greenhouses Venlo-style, high-tech greenhouses that feature a high level of technology to control the indoor climate are the standard for commercial-scale hydroponic CEA. High-tech greenhouses are characterized by computerized controls and automated infrastructure, such as automatic thermal curtains, automatic lighting arrays, and forced-air ventilation systems. These technologies enable a high level of environmental control, though they come at the cost of high energy consumption.

Fig. Eco-Ark Greenhouse at Finn & Roots Farm (Bakersfield, Vermont, USA)

Some large-scale commercial aquaponic farms use this greenhouse typology for their plant production, such as Superior Fresh farms, located in Hixton, Wisconsin, USA (123,000 sf/11,430 m^2), with the aquaculture systems housed in a separate opaque enclosure. Automated supplemental LED lighting and heating enables Superior Fresh farms to cultivate leafy greens year-round despite lack of daylight in the winter, where the natural, frost-free growing season lasts only 4 months. Automated systems for internal climate control allow high-tech greenhouses to be operated anywhere in the world – Blue Smart Farms greenhouse uses an array of sensors to optimize shading during hot Australian summers.

Thanet Earth, the largest greenhouse complex in the UK, is located in the southeast of England. Its five greenhouses cover more than 17 acres (7 hectares) each, growing tomatoes, peppers, and cucumbers using hydroponics (Fig.). This enterprise is powered by a combined heat and power system (CHP) that provides power, heat, and CO_2 for the greenhouses. The CHP system operates very efficiently and channels excess energy to the local district by feeding it into the local power supply grid. In addition, computer-controlled technologies such as energy curtains, high-intensity discharge supplemental lighting, and ventilation regulate the indoor growing conditions.

Rooftop Greenhouses This most recent type includes greenhouses built on top of host buildings, either as retrofits of existing structures or as part of new construction. Due to high land costs, saving space is increasingly important to aquaponic farms in urban contexts. Connecting a greenhouse to an existing building is one strategy for urban farmers looking to revitalize underused space and find a central location in the city. Rooftop greenhouses are already used by commercial-scale hydroponic growers but are a relatively rare enclosure type for aquaponic farms due to the

Fig. Thanet Earth, state of the art greenhouses with combined heat and power provision, (Isle of Thanet in Kent, England, UK)

additional weight of water which can strain existing structures beyond their loading capacity. The few rooftop aquaponic farms that currently exist prioritize lightweight water distribution systems (nutrient film technique or media-based growing rather than deep water culture) and locate their fish tanks on the level below the crop growing space due to relatively decreased demand for natural light.

Two rooftop farms with high-tech aquaponic systems have recently opened in Europe. Both consulted with Efficient City Farming (ECF) farm systems consultants in Berlin. Ecco-jäger Aquaponik Dachfarm in Bad Ragaz, Switzerland sits on top of a distribution center of a family-owned produce company. The Venlo-style rooftop greenhouse (12,900 sf/1200 m^2) is located on a two-story depot building; the fish tanks are installed on the floor below the greenhouse. By growing leafy greens and herbs on their rooftop, Ecco-jäger reduces the need for transportation and can offer produce immediately after harvest. In addition, the farm takes advantage of waste heat generated by its cold storage to heat the greenhouse. BIGH's Ferme Abattoir (21,600 sf/2000 m^2) is a larger version of a similar Venlo-style rooftop greenhouse (Fig.), which occupies the roof of the Foodmet market hall in Brussels, Belgium. These early examples point to further potential to optimize both aquaponic and envelope performance through connecting water, energy, and air flows between farm and host building, known as building-integrated agriculture (BIA). Currently, research is being done on the flagship hydroponic integrated rooftop greenhouse located on the building shared by the Institute of Environmental Science and Technology (ICTA) and the Catalan Institute of Paleontology (ICP) at the Autonomous University of Barcelona (UAB) to dermine

Fig. BIGH Ferme Abattoir with the high-tech greenhouse in the background (Brussels, Belgium)

the benefits of full building integration, although no such example exists in the field of aquaponics to determine the benefits of full building integration, although no such example exists in the field of aquaponics.

Indoor Growing Type

Indoor growing spaces rely exclusively on artificial light for plant production. Often, these growing spaces are highly insulated and clad in an opaque material, originally intended as storage or industrial manufacture rooms. Indoor growing spaces typically have better insulation than greenhouses due to the envelope material, though cannot rely on daylighting or natural heating. The assumption is that this typology is better suited to extreme climates, where temperature swings are of larger concern than lighting (Graamans et al. 2018), though more conclusive research is needed.

Urban Organics operates two commercial-scale indoor growing aquaponic farms within two refurbished breweries in the industrial core of St. Paul, Minnesota, USA. The two farms cultivate leafy greens and herbs in stacked growing beds illuminated by fluorescent grow lights (Fig.). Their second site allows Urban Organics to tap into the brewery infrastructure around an existing aquifer; the aquifer water needs minimal treatment and is supplied at 10 °C to arctic char and rainbow trout

Fig. Urban Organics (St. Paul, Minnesota, USA)

tanks. Using existing structures lowered construction costs for Urban Organics and offered the opportunity to revitalize a struggling area of the city. In an even colder climate, Nutraponics grows leafy greens in a warehouse on a rural parcel 40 km outside Edmonton, Alberta, Canada. Since local produce is highly dependent on seasonal temperature swings, Nutraponics gains a competitive edge in the market by employing LED lighting to accelerate crop growth year-round (Fig.).

Enclosures for Aquaculture

The enclosures for the aquaculture component of aquaponic operations are technically not as demanding as the enclosure design for the hydroponic components since fish do not require sunlight to thrive. Nevertheless, control over indoor growing conditions enables farmers to optimize growth, reduce stress, and draw up precise schedules for fish production which gives their stock a competitive edge in the market (Bregnballe 2015). Aquaculture space enclosures are mainly required to keep water temperatures stable. Fish tanks should be able to support comfortable water temperature ranges for specific fish species, warm-water fish 75–86°F (24–30°C) and cold-water fish 54–74°F (12–23°C) (Alsanius et al. 2017). Water and room temperature can be controlled most efficiently if fish tanks are housed in well-

Fig. Nutraponics (Sherwood Park, Alberta, Canada)

insulated space with few windows to minimize solar gains during the summer months and temperature losses when the outside temperature drops (Pattillo 2017) as demonstrated in the set-up of the INAPRO enclosure. The large volume of water required for fish cultivation needs to be considered from an architectural perspective, as it carries consequences for structural and conditioning systems within a building.

Assessing Enclosure Typologies and Possible Applications

The actual performance of aquaponic farms depends on many case-specific factors. Some preliminary conclusions about enclosure typologies' advantages, challenges, and possible applications can be drawn from the comparison of a relatively small set of case studies. An empirical study of a more significant number of existing case studies will be needed to establish a correlation between enclosure type, geographic location, and commercial success.

Medium-tech greenhouses offer a commercially-feasible option for aquaponic operations only in temperate climates with mild winters and moderate summers, due to their limited environmental control capability. In locations that do not require much heating and cooling, farms using this greenhouse typology can operate in a resource-efficient manner with lower upfront investment for their enclosure. These farms usually operate on a lower budget and include the fish tanks in the same greenhouse, which limits their selection of fish species to those with a large temperature tolerance and draws their commercial focus to the production of lettuce, leafy greens, and herbs.

Passive solar greenhouses rely on passive systems, specifically the use of thermal mass, to control the indoor climate. The use of this typology for aquaponic systems is advantageous since the large volume of water in the fish tanks provides additional thermal mass. Due to their energy efficiency, they are often used in northern latitudes where conventional greenhouses would require a high level of supplemental heating. However, operating any greenhouse in those regions relies on the use of supplemental lighting due to low light levels and short daylight hours during the winter season. Although passive solar greenhouses in Europe and North America are currently used on a small experimental scale, the more general successful application of these single-slope, energy-efficient greenhouses on 1.83 million acres (0.74 million hectares) of farmland in China shows that this typology can be successfully implemented on a large scale (Gao et al. 2010).

High-tech greenhouses, especially large Venlo-style, gutter-connected systems, are the industry standard for commercial hydroponic production. The largest well-funded commercial aquaponic farms use this typology for their hydroponic growing systems in conjunction with a separate enclosure for their aquaculture infrastructure. This setup guarantees the highest level of environmental control as well as crop and fish productivity. Technically, this type of greenhouse can be operated anywhere, as long as the revenue produced pays for the high energy and operation costs in extreme climates. However, this type of operation may not be environmentally sensitive in some northern latitudes due to the extensive need for heating and supplemental lighting. The exact environmental footprint of a high-tech greenhouse can only be assessed on a per-project basis and depends mostly on the quality of energy sources used for supplemental heat and light.

Most *rooftop greenhouses* are Venlo-style high-tech greenhouses constructed on rooftops. Whilst similar benefits and challenges apply, the construction of rooftop greenhouses is even more expensive than that of regular high-tech greenhouses, primarily due to building codes and architectural requirements. The structural system of rooftop greenhouses is often over-dimensioned to comply with building codes for commercial office buildings, which are stricter than building code requirements for agricultural structures. Furthermore, aquaponic operations on rooftops need additional infrastructure to access the roof and comply with fire and egress regulations, which has generated a sprinkler equipped-greenhouse in a recent example (Proksch 2017). The most promising application of rooftop greenhouses is on top of host buildings in urban centers. Urban roofs often offer ample access to sunlight, which greenhouses require to function effectively – a resource that is usually lacking, or at least is not consistent due to shadowing, at ground level in dense urban areas (Ackerman 2012). If purposefully designed, host buildings can offer other resources such as exhaust heat and CO_2 that can make the operation of a rooftop aquaponic farm more feasible. This type of integration with the host building can generate energy and environmental synergies that improve the performance of both greenhouse and host building.

Fig. INAPRO aquaponics enclosure with two sections, opaque for fish and greenhouse for plants (Murcia, Spain)

Indoor growing spaces depend entirely on artificial lighting and active control systems for heating, cooling, and ventilation, which results in a high level of energy consumption, environmental footprint, and operation cost. This typology is most applicable in areas with cold winters and short growing seasons, where the natural exposure to sunlight and heat gain is low and extensive supplementation is needed to operate a commercial aquaponics greenhouse. The use of an opaque enclosure allows high levels of insulation, which reduces heat loss during winter months and provides autonomy from external temperature swings. Besides its dependence on electrical lighting, indoor growing exceeds the productivity of greenhouses as measured in other resources, such as water, CO_2, and land area (Graamans et al. 2018). Additionally, the production per unit of land area can be much higher through the use of stacked growing systems. Regarding the urban integration of aquaponics in cities, indoor grow spaces allow for the adaptive reuse of industrial buildings and warehouses, which can reduce the up-front cost for the construction of the enclosure and support the integration of aquaponic farms in underserved neighborhoods.

The Innovative Aquaponics for Professional Applications (INAPRO, 2018) project set-up included the comparison of the same state of the art aquaponic system and greenhouse technology, across a number of sites in Germany, Belgium, and Spain. The aquaponics system located in China was housed in a passive solar greenhouse. The INAPRO aquaponics facilities in Europe utilized a glass-clad greenhouse type for plant production and an industrial type shed component for fish tanks and filtration units (Fig.). The INAPRO project demonstrates that greenhouse technologies need to be adapted and chosen to suit local climate conditions. The Spanish INAPRO team found, that the selected enclosure was well suited for the cooler northern Europe regions, but not the warmer, Mediterranean regions in southern Europe. This observation highlights the importance of more research on the performance of greenhouses typologies to advance the field of commercial aquaponics operations.

While the comparison of the different typologies reveals certain performance patterns between typology, location, and investment (Table), for a comprehen-sive understanding of farm performance and environmental impact, a more robust system for the analysis and design of farm enclosures is needed.

Impact Assessment as a Design Framework

The growth of aquaponics and generalized claims that aquaponics is more sustainable than other forms of food production has stimulated discussion and research into how sustainable these systems actually are. Life cycle assessment (LCA) is one key quantification method that can be used to analyze sustainability in both agriculture and the built environments by evaluating environmental impacts of products throughout their lifespan. For a building, an LCA can be divided into two types of impact – *embodied* impact which includes material extraction, manufacture, construction, demolition and disposal/reuse of said materials, and *operational* impact which refers to building systems maintenance (Simonen 2014). Similarly, conducting an assessment of an agricultural product can be also divided into the *structural* impact of the building envelope and system infrastructure, *production* impact associated with continuous cultivation and *post-harvest* impact of packaging, storage, and distribution (Payen et al. 2015). Conducting an LCA of an aquaponic farm requires the simultaneous understanding of both building and agricultural impacts since there is an overlap in the envelope's *operational* phase with a crop's *production* phase. The way a building operates its heating, cooling, and lighting systems directly influences the cultivation of the crop; conversely, different types of crops require different environmental conditions. Numerous studies exist comparing LCA results for different building types situated in different contexts (Zabalza Bribián et al. 2009). Similarly, LCA has been used by the agricultural sector to compare efficiencies for different crops and cultivation systems (He et al. 2016; Payen et al. 2015). Evaluating the performance of controlled environment agriculture and aquaponics in particular requires a skillful integration of the two methodologies into one assessment (Sanyé-Mengual 2015).

The proposed aquaponic farm LCA framework (Fig.) is intentionally broad to capture a wide range of farm typologies found in the field. In order to apply the results of LCA to existing farms, factors such as climate and economic data must be included to validate environmental assessment (Goldstein et al. 2016;Rothwell et al. 2016). The following section discusses a collection of aquaponic farm enclosure design strategies based on the LCA inventory of aquaponic farms that synthesizes

Table Comparison of controlled environment agriculture typologies

CEA type	Benefits	Challenges	Cost and revenue[a]
Medium-tech greenhouses	Relies almost entirely on solar energy, low additional energy requirement	Limited environmental control options, susceptible to environmental fluctuations	Lower up-front/ construction cost, (approx. 30–100 $/m^2$)
	Less reliance on non-renewable materials and energy sources	Only applicable to fish species with a large temperature tolerance, (if tanks are in the greenhouse)	
Passive solar greenhouses	Relies on passive systems, uses thermal mass, (including the fish tanks) to buffer temperature swings	Control with passive systems needs more experience and deliberate design	Lower up-front/ construction cost, (approx. 30–100 $/m^2$)
	Low energy consumption, potentially without the need for any fossil fuel	Require supplemental lighting, if located in northern latitudes due to low light levels	
High-tech greenhouses	Highest levels of controls	Relies on active systems for heat, cooling, ventilation and supplemental lighting	High up-front/ construction cost, (approx. 100–200 $/m^2$ and more)
	High productivity with the potential to scale up	High energy consumption and operation cost	
Rooftop greenhouses	Highest levels of controls	Relies on active systems for heat, cooling, ventilation and supplemental lighting	Very high up-front/ construction cost (approx. 300–500 $/m^2$)
	High productivity		
	Potential for energetic and environmental synergies, if integrated with host building	High energy consumption and operation cost	
		Requires code compliance at the level of commercial office buildings	
		Transport of supplies to rooftop is an infrastructural challenge	
Indoor growing spaces	Adaptive reuse of industrial buildings possible	Depends entirely on electrical lighting and active control systems for heating, cooling, and ventilation	Up-front/construction cost can be lower if existing building can be used
	High productivity per unit of footprint though stacked growing systems	High energy consumption and operation cost	Cost depends also on the growing system, stacking multiple levels
	High level of insulation possible		
	Reduced heat loss during winter months		

[a]Based on Proksch (2017)

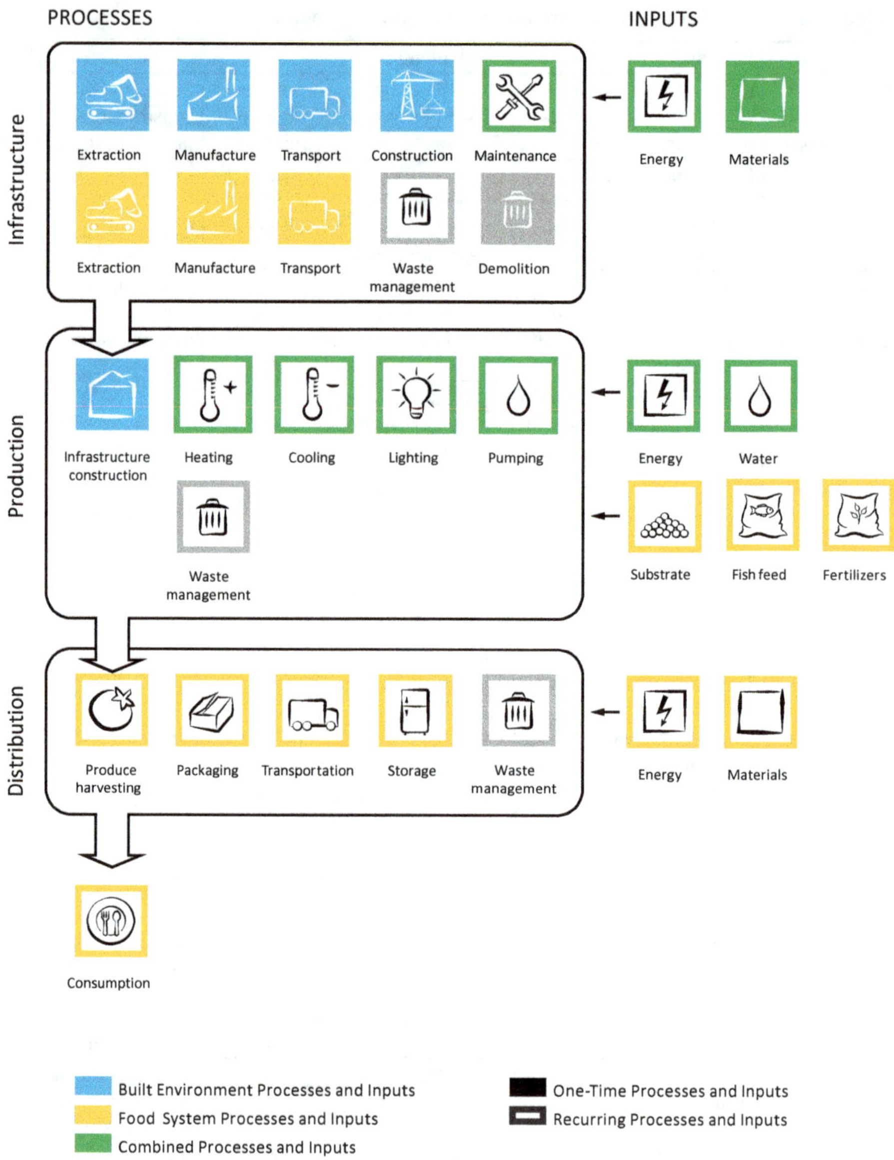

Fig. Example of an integrated LCA process including building and aquaponic system performance. (Based on Sanyé-Mengual et al. 2015).

existing literature with case studies and suggests directions for future work. The unique integration of aquaponic and building-related impacts is of particular interest.

Embodied Impacts: Embodied Energy and Embodied Carbon

Structure Materials and Construction Embodied energy is the calculation of the sum of energy used to extract, refine, process, transport, produce, and assemble a material or product. Embodied carbon is the amount of CO_2 emitted to produce the same material or product. Compared to conventional open-field agricultural operations, the embodied impact of a controlled environment growing system is greater due to increased material extraction and manufacture at the construction stage (Ceron-Palma et al. 2012). For example, in the ICTA-ICP rooftop greenhouse, the structure of the envelope generates 75% more Global Warming Potential (GWP) than a soil-based multi-tunnel greenhouse structure due to the quantity of polycarbonate used in construction (Sanyé-Mengual et al. 2015). Similarly, a building-integrated greenhouse simulation situated in Boston resulted in increased environmental impacts at the construction stage, due to the extraction of iron ores for the manufacture of structural steel (Goldstein 2017). Embodied impacts associated with controlled environment envelopes can be mitigated through smart material use (given that building code adjustments are made to avoid over-sizing structural members) but would nevertheless surpass those of traditional agriculture. Growing food in a constructed envelope will always be more resource-intensive at the beginning compared to simply planting vegetables in an open field, though will also dramatically increase the amount of food that can be produced per area footprint in the same timeframe.

To avoid structure-related environmental impacts, some aquaponic operations make use of existing buildings instead of constructing a new envelope. Urban Organics in St. Paul, Minnesota, USA refurbished two brewery buildings as their indoor growing spaces. In another example of adaptive reuse, The Plant in Chicago, Illinois, USA operates its food incubator and urban farm collective in a 1925 factory building previously used by Peer Foods as a meat-packaging facility (Fig.). Existing insulation and refrigeration equipment were repurposed to control temper-ature fluctuations in the experimental aquaponic facility.

Aquaponic Equipment and Substrate When integrated into buildings, the material choice for aquaponic tanks becomes an important design consideration, since it may limit assembly and transport into the building. For example, polyethylene parts can be assembled on-site using plastic welding, but this is not possible with fiberglass parts (Alsanius et al. 2017). Furthermore, the manufacture of aquaponic system equipment can be a significant contributor to overall environmental impact – for

Fig. The Plant (Chicago, Illinois, USA)

example, glass fiber-reinforced polyester used for the 100 m^3 water tank at the ICTA-ICP rooftop greenhouse is responsible for 10–25% of environmental impact at the manufacturing stage (Fig.). The choice of substrate for plants in an aquaponic system has a weight ramification for the structure of the host building, but also contributes to environmental impact. In a recent study done on aquaponics integrated with living walls, mineral wool, and coconut fiber performed comparably, despite one being compostable and the other being single-use (Khandaker and Kotzen 2018).

Structure and Equipment Maintenance Initial material selection for aquaponic equipment and envelope components determines the long-term upkeep of aquaponic farms. Manufacturing more durable materials such as glass or rigid plastics requires a greater initial investment of environmental resources than plastic films; however, films require replacement more frequently – for example, glass is expected to remain functional for 30+ years, whilst more conventional coated polyethylene film can only last 3–5 years before becoming too opaque (Proksch 2017). Depending on the intended lifespan of an aquaponic system envelope, it may be more advantageous to choose a material with a shorter lifespan, and a lesser manufacturing impact. ETFE film used in the Aquaponic solar greenhouse is a promising compromise between longevity and sustainability, although further research is needed. Standard aquaponic equipment consists of water tanks and piping. Piping for aquaponic systems is often manufactured from PVC, which produces a significant environmental impact in its manufacturing process but does not require replacement for up to 75 years. Some aquaponic suppliers offer bamboo as an organic alternative.

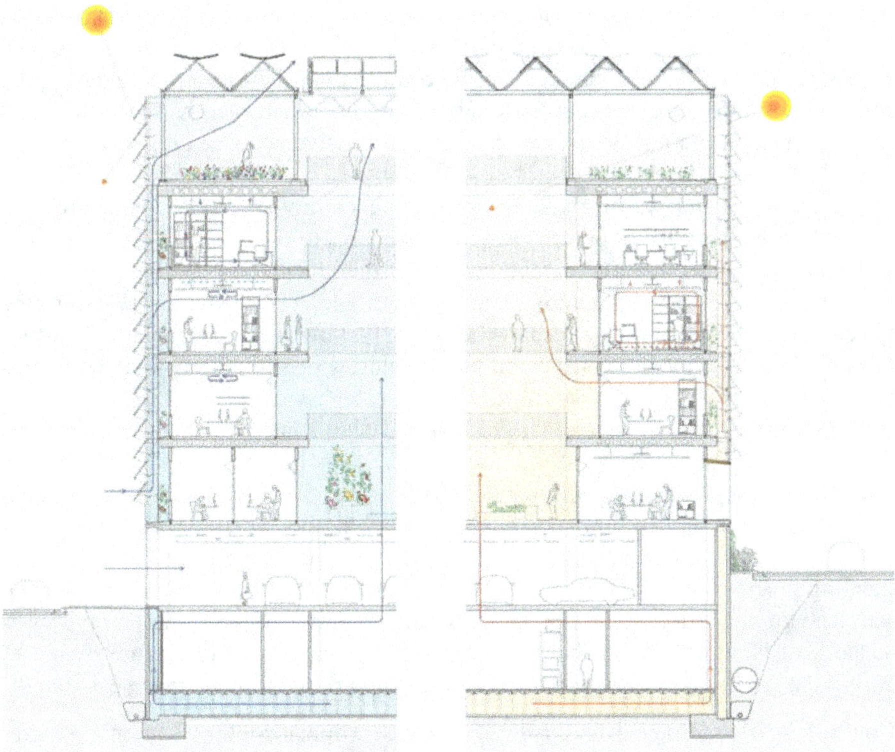

Fig. Building section with rooftop greenhouses by Harquitectes, ICTA-ICP building (Bellaterra, Spain)

Operational Impacts

Energy In 2017, 39% of total energy consumption within the United States corresponded to the building sector (EIA). The agricultural sector accounted for approximately 1.74% of total U.S. primary energy consumption in 2014, relying heavily on indirect expenditures in the form of fertilizers and pesticides (Hitaj and Suttles 2016). Energy efficiency is a well-established field of research within both the built environment and agriculture, often defining the operational impacts of a product, building, or farm in the overall LCA (Mohareb et al. 2017). Integrating building and agricultural energy use can optimize the performance of both (Sanjuan-Delmás et al. 2018).

Heating Energy requirements for heating growing spaces are of particular interest in the northern climates, where extending a naturally short growing season gives building-integrated aquaponic farms a competitive edge in the market (Benis and Ferrão 2018). However, in colder climates, energy consumption by active heating systems is a significant contributor to overall environmental impact – in an assessment of conditioned growing spaces in Boston, Massachusetts, heating costs neutralized the benefits of eliminating food miles in the urban food chain (Benis et al. 2017b; Goldstein 2017). This does not hold true in Mediterranean climates, where climatic conditions are conducive to agriculture and where nearly year-round and conventional greenhouse structures can rely on passive solar heating (Nadal et al. 2017; Rothwell et al. 2016).

In both cold and warm climates, integrating controlled environment growing systems on existing rooftops can provide insulation to the host building – a farm in Montreal, Quebec reports to capture 50% of the greenhouse heating needs from the existing host structure, thereby reducing heating load (Goldstein 2017). Lighting systems can also be partially responsible for satisfying heating demand in interior vertical growing applications such as plant factories or shipping containers (Benis et al. 2017b).

Residual heat capture is another promising design strategy that can optimize the performance of both the host structure and the growing system. Post-occupancy studies of the experimental rooftop greenhouse at the ICTA-ICP in Bellaterra, Spain indicate that the integration of the building with the greenhouse delivered an equivalent carbon savings of 113.8 kg/m^2/year compared to a conventional free-standing greenhouse heated with oil (Nadal et al. 2017). Without intervention from active heating, ventilation and air conditioning (HVAC) systems, the thermal mass of the host laboratory/office building raised the greenhouse temperature by 4.1°C during the coldest months, enabling the cultivation of the tomato crop year-round.

Cooling In Mediterranean and tropical climates, artificial cooling is often a requirement to grow produce year-round. In a rooftop greenhouse simulation, cooling loads represented up to 55% of total farm energy demands in Singapore and in the more temperate climate of Paris, 30% (Benis et al. 2017b). Cooling energy demands are especially high in arid climates, which can benefit the most from cutting conventional transportation costs for perishable produce (Graamans et al. 2018; Ishii et al. 2016). Evaporative cooling, fog cooling, and shading are some strategies for lowering temperatures in aquaponic farms and improving farm performance in terms of yield.

Building-integrated aquaponic systems have the advantage of storing thermal mass in fish tanks to alleviate cooling as well as heating loads. In cases where this mode of passive cooling does not satisfy the cooling demand, evaporative cooling is most commonly used. The Sustainable Harvesters greenhouse produces lettuce for the Houston, Texas, USA area year-round by using a fan and pad cooling system, a subset of evaporative cooling technology. Hot air from outside the envelope first passes through a wet cellulose medium before entering the growing space. As a result, the interior air is cooler and more humid. Evaporative cooling is most effective in dry climates but requires high water use, which may be a limitation to farms in arid areas of the world.

Fog cooling is an alternative strategy. In a fog-cooled greenhouse, plants are periodically misted with water from overhead sprinklers/misters until the space reaches the desired temperature for cultivation. Fog cooling uses less water than evaporative cooling but increases the relative humidity of a growing space. If paired with the right ventilation strategy, fog cooling can be a water-saving technology particularly suited to arid regions (Ishii et al. 2016). Additionally, fog cooling decreases the rate of evapotranspiration in plants, which is critical to optimizing plant metabolism in aquaponic systems (Goddek 2017). The flagship greenhouse of Superior Fresh farms uses a computerized fog-cooling system to maintain cultivation temperatures during the hot season.

Shading devices can also contribute to lowering greenhouse temperatures. Traditionally, the seasonal lime whitewashing of greenhouses was used to reduce solar radiation levels during the hottest months (*Controlled Environment Agriculture* 1973). However, shading can be integrated with other building functions. A promising shading strategy is using semi-transparent photovoltaic modules to simultaneously cool the space and produce energy (Hassanien and Ming 2017). The Aquaponic solar greenhouse combines its photovoltaic array with shading functionality; it uses rotating aluminium panels as shading devices that operate as solar collectors with the help of mounted photovoltaic cells. The integrated photovoltaic system then transforms excess solar radiation into electrical energy.

Lighting The main advantage of greenhouses over indoor growing spaces is their ability to capitalize on daylight to facilitate photosynthesis. However, farms in extreme climates may find that satisfying heating or cooling loads for a transparent envelope is not financially feasible; in this case, farmers may choose to cultivate crop in indoor growing spaces with an insulated envelope (Graamans et al. 2018). Aquaponic farms that operate in indoor growing spaces rely on efficient electrical lighting to produce crops.

Many advances in contemporary farm lighting originated in Japanese plant factories, used to optimize plant yields in dense hydroponic systems by replacing sunlight with engineered light wavelengths (Kozai et al. 2015). Currently, LED lighting is the most popular choice for electrical horticultural lighting systems. They are 80% more efficient than high-intensity discharge lamps and 30% more efficient than their fluorescent counterparts (Proksch 2017). LED lighting continues to be investigated to optimize energy efficiency and crop yield (Zhang et al. 2017). Large-scale greenhouses like Superior Fresh, Wisconsin, USA rely on computerized, supplemental lighting regimes to extend the photosynthesis period of its crop in northern latitudes.

Energy Generation Constrained by the same factors as all CEA, the energy management of an aquaponic farm depends on exterior climate, crop selection, the production system, and structure design (Graamans et al. 2018). Growing produce through aquaponics is not inherently sustainable if not managed properly – all of the factors above can affect energy efficiency for the better or worse (Buehler and Junge 2016). In many cases, CEA is more energy-intensive than conventional open-field agriculture; however, higher energy expenditures may be justified if the way we source energy shifts toward renewable sources and efficient strategies for heating, cooling, and lighting are incorporated into the design of the farm.

Photovoltaic (PV) power generation can play an important part in offsetting operational impacts for controlled environment aquaponics, reducing environmental strain. In an example of a high-tech greenhouse in Australia, using energy from a PV array caused a 50% reduction in lifecycle greenhouse gas emissions compared to the conventional grid scenario (Rothwell et al. 2016). Renewable energy generation can be combined with aquaponic farms, space permitting – for example, the Lucky Clays Fresh aquaponic greenhouse on a rural farm in North Carolina runs on energy generated by wind turbines and photovoltaic panels that are situated elsewhere on the owner's land parcel.

Water Water use efficiency has been often cited as a major benefit of CEA and hydroponic systems (Despommier 2013; Specht et al. 2014). Aquaponic systems are even better suited to increase water efficiency – where 1 kg of fish produced in a conventional aquaculture system requires between 2500 and 375,000 L, the same amount of fish raised in an aquaponic system requires less than 100 L (Goddek et al. 2015). Rainwater capture and greywater reuse have been proposed as two strategies to offset the watershed impacts of operating a hydroponic or aquaponic farm even further. At the existing ICTA-ICP greenhouse, 80–90% of the water needs for the production of tomatoes in an aggregate hydroponic system were covered by rainwater capture within a year of operation (Sanjuan-Delmás et al. 2018). However, the ability of rainwater capture to meet crop demand depends on the climatic context. In a study evaluating the viability of rooftop greenhouse production on existing retail parks in eight cities around the world, seven met crop self-sufficiency through rainwater capture – only Berlin did not (Sanyé-Mengual et al. 2018).

Some existing CEA facilities already reuse greywater to improve efficiency (Benke and Tomkins 2017). However, greywater reuse in an urban context is currently limited due to lacking regulatory support and currently-lacking research on the health risks of using greywater in agriculture. A pilot of greywater reuse, the Maison Productive in Montréal collects greywater from household uses to supplement its rainwater collection to irrigate gardens and a communal greenhouse for food production that nine residential units share (Thomaier et al. 2015). With further advances in policy on the treatment of greywater, building-integrated aquaponics can tap into the existing water cycle instead of relying on municipal sources.

From an architectural standpoint, water distribution in an aquaponic system is likely to present a structural challenge. Aquaponic fish tanks weigh more than hydroponic grow beds and may limit what types of structures are feasible for retrofitting an aquaponic farm. The growing medium also requires consideration – deep water culture (DWC) systems require a large and heavy volume of water, whilst nutrient film technique (NFT) systems are lightweight but expensive to manufacture (Goddek et al. 2015).

Nutrients Compared to conventional open-field farming, CEA reduces the need for fertilizers and pesticides, as the farmer can physically separate the crop from harsh external conditions (Benke and Tomkins 2017). However, due to the density of an aquaponic system, plant or fish diseases can spread quickly if a pathogen infiltrates the space. Preventative options such as the use of predator insects or tight environmental control measures such as a "buffer" entryways can avert this risk (Goddek et al. 2015).

The integration of different fish and crop nutrient needs is a challenge in single-recirculating aquaponic systems (Alsanius et al. 2017). Generally, plants require higher nitrogen concentrations than fish can withstand and careful crop and fish selection can match nutrient requirements to optimize yields, but is still difficult to achieve. Decoupled systems (DRAPS) have been proposed to separate the aquaculture water cycle from the hydroponic one to achieve desired nutrient concentrations, but is not yet commonly applied in commercial farms (Suhl et al. 2016). Urban Organics based in St. Paul, Minnesota, USA chose to develop a DRAPS system for their second farm to optimize both crop and fish yields and avoid crop loss in case of nutrient imbalances within fish tanks. ECF Farm in Berlin, Germany, and Superior Fresh farms in Wisconsin, USA also operate decoupled systems to optimize fish and plant growth.

Alternatively, aquaponic nutrient cycles can be optimized through the introduction of an anaerobic reactor to transform solid fish waste into plant-digestible phosphorus (Goddek et al. 2016). Currently, The Plant in Chicago, USA is planning to operate an anaerobic digester which may play a part in optimizing nutrient cycles for crop growth. The mechanical system requirements for DRAPS and anaerobic digestion will influence the performance as well as the spatial layout of an aquaponic farm.

End-of-Life Impacts

Materials Waste Management A theoretical advantage of CEA over open-field farming is the ability to control materials waste runoff, preventing leaching (Despommier 2013; Gould and Caplow 2012). A tight envelope can play a role in efficient materials waste management. One pathway of recycling organic waste matter to improve building performance is the use of plant stalks for the production of insulating biochar, although this research is in early stages (Llorach-Massana et al. 2017). Additionally, considering the incorporation of waste management components such as a filtration bed, an anaerobic digester or a heat recovery ventilator into the enclosure design at an early stage can close energy, nutrient, and water loops for the farm.

Distribution Chains Packaging has been a hotspot in various farm LCAs assessing the impact of production. It is responsible for as much as 45% of the total impact for a tomato in Bologna, Italy, and is the largest contributor to the environmental impacts of indoor hydroponic systems in Stockholm, Sweden (Molin and Martin 2018b; Orsini et al. 2017; Rothwell et al. 2016). Siting aquaponic farms close to consumers can reduce the need for packaging, storage, and transport as with other forms of urban agriculture, if local retailers and distributors collaborate with farmers (Specht et al. 2014). Unfortunately, due to consumer acceptance, most large-scale retailers currently require standard plastic packaging for aquaponic produce to be sold alongside conventional brands - – therefore, selecting a site close to a consumer market for controlled environment aquaponics does not guarantee significant changes in the overall performance of the farm.

Reduced transportation, or food-miles, is often cited in the literature as a major advantage of urban agriculture (Benke and Tomkins 2017; Despommier 2013; Sanjuan-Delmás et al. 2018). However, it is important to note that the relative contribution of shortened transportation chains varies on a case-by-case basis. In Singapore, where nearly all food has to be imported from neighboring countries, cutting transportation chains makes sense financially and in terms of environmental impact (Astee and Kishnani 2010). The same cannot be said for Spain, where the conventional supply chain of tomatoes from farm to city is already short (Sanjuan-Delmás et al. 2018). Cities with the longest supply chains can benefit from localized food production, but the benefits of cutting transportation must be weighed against operational and embodied impacts. In the case of Boston, the benefits of reduced transportation were entirely negated by the impact of heating and operating a greenhouse inside the city (Goldstein 2017). Despite long conventional food supply chains, transportation impacts were similarly insignificant in the bigger picture of CEA performance in Stockholm (Molin and Martin 2018a).

Consumption and Diet Aquaponic farms in cities can alter urban diets, which play a significant role in the environmental impact of food consumption (Benis and Ferrão 2017). Meat consumption via the conventional chain produces the largest share of the current environmental footprint and seeking protein alternatives has the potential for a larger impact than the widespread implementation of urban agriculture (Goldstein 2017). Since aquaponics produces fish as well as vegetables, this potential to change protein diets on a large scale should not be ignored in larger assessments of environmental performance.

Integrated Urban Aquaponics

When deliberately designed with respect to environmental impact, aquaponic farms can become part of a resource-efficient urban food system. No aquaponic farm operates in isolation since when crops are harvested and reach the farm gate, they enter a larger socioeconomic food network as fish and produce is distributed to customers. At this stage, the performance of aquaponic farms is no longer confined to the growing system and envelope – economics, marketing, education, and social outreach are also involved. Urban aquaponic farms will need to operate as competitive businesses and good neighbors to be successfully integrated into city life.

Economic Viability

The economic viability of aquaponic farms depends on many contextual factors where both local conventional fish production chains and open-field farming must be matched (Stadler et al. 2017). While aquaponics requires a relatively costly initial investment, it may outperform conventional farming during the production and distribution phase where the design of the recirculating water system reduces water costs, and greatly reduces the need for fertilizers, which usually comprise between 5% and 10% of overall farm costs (Hochmuth and Hanlon 2010). However, estimating the economic viability of aquaponic farms is particularly challenging due to the range of dynamic factors affecting performance including the local price for labor and energy being two examples (Goddek et al. 2015). In an economic analysis of aquaponic farms in the Midwestern United States, labor constituted 49% of all operational costs despite the assumption that only minimum wages would be paid. In reality, the wide range of expertise required to operate an aquaponic system will likely warrant higher wages in an urban farm scenario (Quagrainie et al. 2018).

Site selection and envelope design have a direct relationship to the profitability of an aquaponic farm by affecting operation efficiency and how broad the potential market can be. Aquaponic farms located in urban environments can tap into multiple markets outside agricultural production, where many aquaponic farms offer tours, workshops, design consulting services, and supply backyard aquaponic systems for hobbyists. Integrating agriculture with other types of spaces within urban environments can contribute to the financial health of aquaponic farms. The ECF aquaponic farm is located on the work yard of the industrial landmark building Malzfabrik, Berlin, Germany, which operates a cultural center and houses work spaces for artists and designers.

Accessibility and Food Security

Urban agriculture is often cited as a strategy to provide fresh food for underserved communities located in food deserts, yet few commercial urban farms target this demographic, proving that commercial-scale urban agriculture can be just as exclusionary as conventional supply chains (Gould and Caplow 2012; Sanyé-Mengual et al. 2018; Thomaier et al. 2015). Aquaponic farms that use high-tech infrastructure try to redeem their high investments by achieving premium prices in urban markets, though aquaponics can also stem from grassroots and hobbyist applications. Aquaponics may also have the potential to increase food security for urban residents. This is evidenced in the lasting legacy of Growing Power, a non-profit organization that until recently, ran an urban farm in Milwaukee, Wisconsin, USA started by Will Allen in 1993. Many current aquaponic farmers attended Growing Power's workshops, in which Allen championed an aquaponic model that gives back to the surrounding community by means of community-supported agriculture boxes and classes. Initiated by Growing Power's educational programs, other aquaponic non-profit organizations have taken up to the torch such as Dre Taylor with Nile Valley Aquaponics in Kansas City, Kansas, USA. This farm aims to provide 100,000 pounds (45,400 kg) of local produce to the surrounding community in an award-winning new campus for the expanding farm (Fig.).

Fig. Proposed Nile Valley Aquaponics campus (Kansas City, Kansas, USA) by HOK Architects

Education and Job Training

Aquaponics can be used as an educational tool to promote systems thinking and environmental mindfulness (Junge et al. 2014; Specht et al. 2014). In urban applications, aquaponic systems could be used to raise awareness of ecological cycles much like existing soil-based farms (Kulak et al. 2013). The Greenhouse Project in New York City translates this into a new approach to science education in public schools. The Greenhouse Project aims to build 100 rooftop greenhouses on public schools as science classrooms. These greenhouses, customized for their dual mission of growing and learning, all include an aquaponic system. However, aquaponic systems also require greater collaboration between existing academic disciplines in order to move forward in this new multidisciplinary academic field (Goddek et al. 2015). The collaboration of aquaculture and horticulture specialists, engineers, business strategists, and built environment professionals amongst many others is necessary to turn aquaponics into an important contributor to sustainable urban development.

Conclusions

There is an array of criteria that contribute to the performance of each farm and their number grows with the number of disciplines involved in this the interdisciplinary field of aquaponics. Of note is an earlier study that has provided a definition of aquaponics and a classification of the types of aquaponics based on size and system

(Palm et al. 2018). Many criteria for the analysis of the enclosure type identified in this study stem from immediate farm context – local climate, the quality of the built environment context, energy sourcing practices, costs, market, and local regulatory frameworks. An aquaponic greenhouse in a rural context performs differently than one in a city, just as farms in arid climates do not share the same requirements as their counterparts in colder areas. In general, greenhouses classified as medium-tech and passive solar offer a lower cost, environmentally sustainable enclosure option, currently only used by smaller aquaponic operations. However, due to their intentionally limited level of technical environmental controls, they only perform well in specific climate zones. In comparison, high-tech and rooftop greenhouses can be technically implemented anywhere, though in extreme climate conditions they generate high operational costs and larger environmental footprints. Recent case studies show that indoor growing facilities can be financially feasible, but due to their exclusive reliance on electrical lighting, their resource use efficiency and environmental footprint are of concern. Further research is needed to establish the relationship of specific aquaponic farms and their enclosures to existing resource networks. This work can help connect aquaponics to research done on urban metabolism.

Other criteria determining farm typology and performance are internal. These include environmental control levels, crop and fish selection, aquaponic system type and scale and enclosure type and scale. Taking on an integrated LCA approach, the relationship between all factors have to be assessed throughout the lifespan of the farm, from cradle to grave. Life cycle assessment of aquaponic farms must include both building impacts and growing system impacts since there is overlap in the farm operation phase. A series of promising strategies in heating, cooling, lighting, and material design can improve overall farm efficiency throughout the entire lifespan of the farm. Beyond accounting for environmental impact, LCA can become a design framework for horticulture experts, aquaculture specialists, architects, and investors.

Continuing to survey existing commercial aquaponic farms is important to validate LCA models, identify strategies, and cataloguing aquaponic operations emerging on a larger scale. Combining modeling with case study research on controlled environment aquaponics has the potential to connect aquaponics to the larger scope of urban sustainability.

Acknowledgments The authors of this study acknowledge the financial support of the National Science Foundation (NSF) under the umbrella of the Sustainable Urbanization Global Initiative (SUGI) Food Water Energy Nexus and the support of all CITYFOOD project partners for providing ideas and inspiration.

References

Ackerman K (2012) The potential for urban agriculture in New York City: growing capacity, food security and green infrastructure report. Columbia University Urban Design Lab, New York

Alsanius BW, Khalil S, Morgenstern R (2017) Rooftop aquaponics. In: Rooftop urban agriculture, urban agriculture. Springer, Cham, pp 103–112. https://doi.org/10.1007/978-3-319-57720-3_7

Astee LY, Kishnani NT (2010) Building integrated agriculture: utilising rooftops for sustainable food crop cultivation in Singapore. J Green Build 5:105–113. https://doi.org/10.3992/jgb.5.2.105

Benis K, Ferrão P (2017) Potential mitigation of the environmental impacts of food systems through urban and peri-urban agriculture (UPA) – a life cycle assessment approach. J Clean Prod 140:784–795. https://doi.org/10.1016/j.jclepro.2016.05.176

Benis K, Ferrão P (2018) Commercial farming within the urban built environment – taking stock of an evolving field in northern countries. Glob Food Sec 17:30–37. https://doi.org/10.1016/j.gfs.2018.03.005

Benis K, Reinhart C, Ferrão P (2017a) Development of a simulation-based decision support workflow for the implementation of building-integrated agriculture (BIA) in urban contexts. J Clean Prod 147:589–602. https://doi.org/10.1016/j.jclepro.2017.01.130

Benis K, Reinhart C, Ferrão P (2017b) Building-integrated agriculture (BIA) in urban contexts: testing a simulation-based decision support workflow. Presented at the Building Simulation 2017, San Francisco, USA, p 10. https://doi.org/10.26868/25222708.2017.479

Benke K, Tomkins B (2017) Future food-production systems: vertical farming and controlled-environment agriculture. Sustain Sci Pract Policy 13:13–26. https://doi.org/10.1080/15487733.2017.1394054

Bregnballe J (2015) A guide to recirculation aquaculture: an introduction to the new environmentally friendly and highly productive closed fish farming systems. Food and Agriculture Organization of the United Nations: Eurofish, Copenhagen

Buehler D, Junge R (2016) Global trends and current status of commercial urban rooftop farming. Sustainability 8:1108. https://doi.org/10.3390/su8111108

Ceron-Palma I, Sanyé-Mengual E, Oliver-Solà J, Rieradevall J (2012) Barriers and opportunities regarding the implementation of rooftop eco.greenhouses (RTEG) in mediterranean cities of Europe. J Urban Technol 19:87–103. https://doi.org/10.1080/10630732.2012.717685

Controlled Environment Agriculture (1973) A global review of greenhouse food production (No. 89), Economic Research Service. U.S. Department of Agriculture

de Graaf PA (2012) Room for urban agriculture in Rotterdam: defining the spatial opportunities for urban agriculture within the industrialised city. In: Sustainable food planning: evolving theory and practice. Wageningen Academic Publishers, Wageningen, pp 533–546. https://doi.org/10.3920/978-90-8686-187-3_42

De La Salle JM, Holland M (2010) Agricultural urbanism. Green Frigate Books

Despommier D (2013) Farming up the city: the rise of urban vertical farms. Trends Biotechnol 31:388–389. https://doi.org/10.1016/j.tibtech.2013.03.008

dos Santos MJPL (2016) Smart cities and urban areas–aquaponics as innovative urban agriculture. Urban For Urban Green 20:402–406. https://doi.org/10.1016/j.ufug.2016.10.004

EU Aquaponics Hub (2017) COST Action FA1305, Aquaponics map (Cost FA1305), https://www.google.com/maps/d/u/0/viewer?ll=50.77598474809961%2C12.62131196967971&z=4&mid=1bjUUbCtUfE_BCgaAf7AbmxyCpT0

Fang Y, Hu Z, Zou Y, Zhang J, Zhu Z, Zhang J, Nie L (2017) Improving nitrogen utilization efficiency of aquaponics by introducing algal-bacterial consortia. Bioresour Technol 245:358–364. https://doi.org/10.1016/j.biortech.2017.08.116

Gao L-H, Qu M, Ren H-Z, Sui X-L, Chen Q-Y, Zhang Z-X (2010) Structure, function, application, and ecological benefit of a single-slope, energy-efficient solar greenhouse in China. HortTechnology 20:626–631

Goddek S (2017) Opportunities and challenges of multi-loop aquaponic systems. Wageningen University, Wageningen

Goddek S, Delaide B, Mankasingh U, Ragnarsdottir KV, Jijakli H, Thorarinsdottir R (2015) Challenges of sustainable and commercial Aquaponics. Sustainability 7:4199–4224. https://doi.org/10.3390/su7044199

Goddek S, Schmautz Z, Scott B, Delaide B, Keesman KJ, Wuertz S, Junge R (2016) The effect of anaerobic and aerobic fish sludge supernatant on hydroponic lettuce. Agronomy 6:37. https://doi.org/10.3390/agronomy6020037

Goldstein BP (2017) Assessing the edible city: environmental implications of urban agriculture in the Northeast United States. Technical University of Denmark, Lyngby

Goldstein B, Hauschild M, Fernández J, Birkved M (2016) Testing the environmental performance of urban agriculture as a food supply in northern climates. J Clean Prod 135:984–994. https://doi.org/10.1016/j.jclepro.2016.07.004

Gould D, Caplow T (2012) Building-integrated agriculture: a new approach to food production. In: Metropolitan sustainability: understanding and improving the urban environment. Woodhead Publishing Limited, Cambridge, pp 147–170

Graamans L, Baeza E, van den Dobbelsteen A, Tsafaras I, Stanghellini C (2018) Plant factories versus greenhouses: comparison of resource use efficiency. Agric Syst 160:31–43. https://doi.org/10.1016/j.agsy.2017.11.003

Hassanien RHE, Ming L (2017) Influences of greenhouse-integrated semi-transparent photovoltaics on microclimate and lettuce growth. Int J Agric Biol Eng 10:11–22. https://doi.org/10.25165/ijabe.v10i6.3407

He X, Qiao Y, Liu Y, Dendler L, Yin C, Martin F (2016) Environmental impact assessment of organic and conventional tomato production in urban greenhouses of Beijing city, China. J Clean Prod 134:251–258. https://doi.org/10.1016/j.jclepro.2015.12.004

Hitaj C, Suttles S (2016) Trends in U.S. agriculture's consumption and production of energy: renewable power, shale energy, and cellulosic biomass, Economic information bulletin, no. 159. USDA/Economic Research Service, Washington, DC

Hochmuth GJ, Hanlon EA (2010) Commercial vegetable fertilization principles 17

INAPRO - Innovative Aquaponics for Professional Applications (2018). http://inapro-project.edu

Ishii M, Sase S, Moriyama H, Okushima L, Ikeguchi A, Hayashi M, Kurata K, Kubota C, Kacira M, Giacomelli GA (2016) Controlled environment agriculture for effective plant production systems in a semiarid greenhouse. JARQ 50:101–113. https://doi.org/10.6090/jarq.50.101

Junge R, Wilhelm S, Hofstetter U (2014) Aquaponic in classrooms as a tool to promote system thinking. In: Transmission of innovations, knowledge and practical experience into everyday practice. Presented at the Conference VIVUS – on agriculture, environmentalism, horticulture and floristics, food production and processing and nutrition, Naklo, Slovenia, p 11

Junge R, König B, Villarroel M, Komives T, Jijakli MH (2017) Strategic points in aquaponics. Water 9:182. https://doi.org/10.3390/w9030182

Khandaker M, Kotzen B (2018) The potential for combining living wall and vertical farming systems with aquaponics with special emphasis on substrates. Aquac Res 49:1454–1468. https://doi.org/10.1111/are.13601

König B, Junge R, Bittsanszky A, Villarroel M, Komives T (2016) On the sustainability of aquaponics. Ecocycles 2(1):26–32. https://doi.org/10.19040/ecocycles.v2i1.50

Körner O, Gutzmann E, Kledal PR (2017) A dynamic model simulating the symbiotic effects in aquaponic systems. Acta Hortic 1170:309–316. https://doi.org/10.17660/ActaHortic.2017.1170.37

Kozai T, Niu G, Takagaki M (2015) Plant factory: an indoor vertical farming system for efficient quality food production. Academic

Kulak M, Graves A, Chatterton J (2013) Reducing greenhouse gas emissions with urban agriculture: a life cycle assessment perspective. Landsc Urban Plan 111:68–78. https://doi.org/10.1016/j.landurbplan.2012.11.007

Lastiri DR, Geelen C, Cappon HJ, Rijnaarts HHM, Baganz D, Kloas W, Karimanzira D, Keesman KJ (2018) Model-based management strategy for resource efficient design and operation of an aquaponic system. Aquac Eng 83:27. https://doi.org/10.1016/j.aquaeng.2018.07.001

Llorach-Massana P, Lopez-Capel E, Peña J, Rieradevall J, Montero JI, Puy N (2017) Technical feasibility and carbon footprint of biochar co-production with tomato plant residue. Waste Manag 67:121–130. https://doi.org/10.1016/j.wasman.2017.05.021

Maucieri C, Forchino AA, Nicoletto C, Junge R, Pastres R, Sambo P, Borin M (2018) Life cycle assessment of a micro aquaponic system for educational purposes built using recovered material. J Clean Prod 172:3119–3127. https://doi.org/10.1016/j.jclepro.2017.11.097

Mohareb E, Heller M, Novak P, Goldstein B, Fonoll X, Raskin L (2017) Considerations for reducing food system energy demand while scaling up urban agriculture. Environ Res Lett 12:125004. https://doi.org/10.1088/1748-9326/aa889b

Molin E, Martin M (2018a) Assessing the energy and environmental performance of vertical hydroponic farming (No. C 299). ICL Swedish Environmental Research Institute, ICL Swedish Environmental Research Institute

Molin E, Martin M (2018b) Reviewing the energy and environmental performance of vertical farming systems in urban environments (No. C 298). ICL Swedish Environmental Research Institute, ICL Swedish Environmental Research Institute

Monsees H, Kloas W, Wuertz S (2017) Decoupled systems on trial: eliminating bottlenecks to improve aquaponic processes. PLoS One 12:e0183056. https://doi.org/10.1371/journal.pone.0183056

Nadal A, Llorach-Massana P, Cuerva E, López-Capel E, Montero JI, Josa A, Rieradevall J, Royapoor M (2017) Building-integrated rooftop greenhouses: an energy and environmental assessment in the mediterranean context. Appl Energy 187:338–351. https://doi.org/10.1016/j.apenergy.2016.11.051

Orsini F, Dubbeling M, de Zeeuw H, Prosdocimi Gianquinto GG (2017) Rooftop urban agriculture, Urban agriculture (springer (firm)). Springer, Cham

Palm HW, Knaus U, Appelbaum S, Goddek S, Strauch SM, Vermeulen T, Jijakli M, Kotzen B (2018) Towards commercial aquaponics : a review of systems, designs, scales and nomenclature. Aquac Int 26(3):813–842. ISSN 0967-6120. https://doi.org/10.1007/s10499-018-0249-z

Pattillo DA (2017) An overview of aquaponic systems: aquaculture components (No. 20), NCRAC Technical Bulletins. North Central Regional Aquaculture Center

Payen S, Basset-Mens C, Perret S (2015) LCA of local and imported tomato: an energy and water trade-off. J Clean Prod 87:139–148. https://doi.org/10.1016/j.jclepro.2014.10.007

Pearson LJ, Pearson L, Pearson CJ (2010) Sustainable urban agriculture: stocktake and opportunities. Int J Agric Sustain 8:7–19. https://doi.org/10.3763/ijas.2009.0468

Proksch G (2017) Creating urban agriculture systems: an integrated approach to design. Routledge, New York

Quagrainie KK, Flores RMV, Kim H-J, McClain V (2018) Economic analysis of aquaponics and hydroponics production in the U.S. Midwest. J Appl Aquac 30:1–14. https://doi.org/10.1080/10454438.2017.1414009

Rothwell A, Ridoutt B, Page G, Bellotti W (2016) Environmental performance of local food: trade-offs and implications for climate resilience in a developed city. J Clean Prod 114:420–430. https://doi.org/10.1016/j.jclepro.2015.04.096

Sanjuan-Delmás D, Llorach-Massana P, Nadal A, Ercilla-Montserrat M, Muñoz P, Montero JI, Josa A, Gabarrell X, Rieradevall J (2018) Environmental assessment of an integrated rooftop greenhouse for food production in cities. J Clean Prod 177:326–337. https://doi.org/10.1016/j.jclepro.2017.12.147

Sanyé-Mengual E (2015) Sustainability assessment of urban rooftop farming using an interdisciplinary approach. Universitat Autònoma de Barcelona, Bellaterra

Sanyé-Mengual E, Oliver-Solà J, Montero JI, Rieradevall J (2015) An environmental and economic life cycle assessment of rooftop greenhouse (RTG) implementation in Barcelona, Spain. Assessing new forms of urban agriculture from the greenhouse structure to the final product level. Int J Life Cycle Assess 20:350–366. https://doi.org/10.1007/s11367-014-0836-9

Sanyé-Mengual E, Martinez-Blanco J, Finkbeiner M, Cerdà M, Camargo M, Ometto AR, Velásquez LS, Villada G, Niza S, Pina A, Ferreira G, Oliver-Solà J, Montero JI, Rieradevall J (2018) Urban horticulture in retail parks: environmental assessment of the potential implementation of rooftop greenhouses in European and south American cities. J Clean Prod 172:3081–3091. https://doi.org/10.1016/j.jclepro.2017.11.103

Simonen K (2014) Life cycle assessment. Routledge, London

Specht K, Siebert R, Hartmann I, Freisinger UB, Sawicka M, Werner A, Thomaier S, Henckel D, Walk H, Dierich A (2014) Urban agriculture of the future: an overview of sustainability aspects of food production in and on buildings. Agric Hum Values 31:33–51. https://doi.org/10.1007/s10460-013-9448-4

Stadler MM, Baganz D, Vermeulen T, Keesman KJ (2017) Circular economy and economic viability of aquaponic systems: comparing urban, rural and peri-urban scenarios under Dutch conditions. Acta Hortic 1176:101–114. https://doi.org/10.17660/ActaHortic.2017.1176.14

Stewart ID, Oke TR (2010) Thermal differentiation of local climate zones using temperature observations from urban and rural field sites. Presented at the Ninth symposium on urban environment, Keystone, CO, p 8

Suhl J, Dannehl D, Kloas W, Baganz D, Jobs S, Scheibe G, Schmidt U (2016) Advanced aquaponics: evaluation of intensive tomato production in aquaponics vs. conventional hydroponics. Agric Water Manag 178:335–344. https://doi.org/10.1016/j.agwat.2016.10.013

Thomaier S, Specht K, Henckel D, Dierich A, Siebert R, Freisinger UB, Sawicka M (2015) Farming in and on urban buildings: present practice and specific novelties of zero-acreage farming (ZFarming). Renewable Agric Food Syst 30:43–54. https://doi.org/10.1017/S1742170514000143

Van Woensel L, Archer G, Panades-Estruch L, Vrscaj D, European Parliament, Directorate-General for Parliamentary Research Services (2015) Ten technologies which could change our lives: potential impacts and policy implications: in depth analysis. European Commission/EPRS European Parliamentary Research Service, Brussels

Zabalza Bribián I, Aranda Usón A, Scarpellini S (2009) Life cycle assessment in buildings: state-of-the-art and simplified LCA methodology as a complement for building certification. Build Environ 44:2510–2520. https://doi.org/10.1016/j.buildenv.2009.05.001

Zhang H, Burr J, Zhao F (2017) A comparative life cycle assessment (LCA) of lighting technologies for greenhouse crop production. Journal of Cleaner Production, Towards eco-efficient agriculture and food systems: selected papers addressing the global challenges for food systems, including those presented at the Conference "LCA for Feeding the planet and energy for life" (6–8 October 2015, Stresa & Milan Expo, Italy) 140:705–713. https://doi.org/10.1016/j.jclepro.2016.01.014

Permissions

All chapters in this book were first published in AFPS, by Springer Nature; hereby published with permission under the Creative Commons Attribution License or equivalent. Every chapter published in this book has been scrutinized by our experts. Their significance has been extensively debated. The topics covered herein carry significant findings which will fuel the growth of the discipline. They may even be implemented as practical applications or may be referred to as a beginning point for another development.

The contributors of this book come from diverse backgrounds, making this book a truly international effort. This book will bring forth new frontiers with its revolutionizing research information and detailed analysis of the nascent developments around the world.

We would like to thank all the contributing authors for lending their expertise to make the book truly unique. They have played a crucial role in the development of this book. Without their invaluable contributions this book wouldn't have been possible. They have made vital efforts to compile up to date information on the varied aspects of this subject to make this book a valuable addition to the collection of many professionals and students.

This book was conceptualized with the vision of imparting up-to-date information and advanced data in this field. To ensure the same, a matchless editorial board was set up. Every individual on the board went through rigorous rounds of assessment to prove their worth. After which they invested a large part of their time researching and compiling the most relevant data for our readers.

The editorial board has been involved in producing this book since its inception. They have spent rigorous hours researching and exploring the diverse topics which have resulted in the successful publishing of this book. They have passed on their knowledge of decades through this book. To expedite this challenging task, the publisher supported the team at every step. A small team of assistant editors was also appointed to further simplify the editing procedure and attain best results for the readers.

Apart from the editorial board, the designing team has also invested a significant amount of their time in understanding the subject and creating the most relevant covers. They scrutinized every image to scout for the most suitable representation of the subject and create an appropriate cover for the book.

The publishing team has been an ardent support to the editorial, designing and production team. Their endless efforts to recruit the best for this project, has resulted in the accomplishment of this book. They are a veteran in the field of academics and their pool of knowledge is as vast as their experience in printing. Their expertise and guidance has proved useful at every step. Their uncompromising quality standards have made this book an exceptional effort. Their encouragement from time to time has been an inspiration for everyone.

The publisher and the editorial board hope that this book will prove to be a valuable piece of knowledge for researchers, students, practitioners and scholars across the globe.

List of Contributors

T. Reinhardt
GFA Consulting Group, Berlin, Germany

K. Hoevenaars
AquaBioTech Group, Mosta, Malta

A. Joyce
Department of Marine Science, University of Gothenburg, Gothenburg, Sweden

B. E. Mikkelsen
Integrated Food Studies, Aalborg University Copenhagen, Copenhagen, Denmark

C. M. Bosire
Department of Learning & Philosophy, Aalborg University, Copenhagen, Denmark

H. Yavuzcan Yildiz
Department of Fisheries and Aquaculture, Ankara University, Ankara, Turkey

V. Radosavljevic
Department of Fish Diseases, National Reference Laboratory for Fish Diseases, Institute of Veterinary Medicine of Serbia, Belgrade, Serbia

G. Parisi
Department of Agriculture, Food, Environment and Forestry (DAGRI), Animal Science Section, University of Florence, Florence, Italy

A. Cvetkovikj
Department of Parasitology and Parasitic diseases, Faculty of Veterinary Medicine, Ss. Cyril and Methodius University in Skopje, Skopje, Republic of Macedonia
Department of Fisheries, Institute of Animal Science, Ss. Cyril and Methodius University in Skopje, Skopje, Republic of Macedonia

L. Robaina
Aquaculture Research Group (GIA), Ecoaqua Institute, University of Las Palmas de Gran Canaria, Telde, Gran Canaria, Spain

J. Pirhonen
Department of Biological and Environmental Science, University of Jyväskylä, Jyväskylä, Finland

E. Mente
Department of Ichthyology and Aquatic Environment, University of Thessaly, Volos, Greece

J. Sánchez
Department of Physiology, Faculty of Biology, Regional Campus of International Excellence "Campus Mare Nostrum", University of Murcia, Murcia, Spain

N. Goosen
Department of Process Engineering, Stellenbosch University, Stellenbosch, South Africa

R. Morgenstern, I. Schröter and M. Mergenthaler
Department of Agriculture, University of Applied Sciences of South Westphalia, Soest, Germany

S. Hüttel
Institute of Food and Resource Economics, Chair of Production Economics, Faculty of Agriculture, Rheinische Friedrichs-Wilhelms Universität Bonn, Bonn, Germany

M. Leyer
Institute of Business Administration, Rostock University, Rostock, Germany

J. Gott
Geography and Environment, University of Southampton, Southampton, UK

M. Turnšek
Faculty of Tourism, University of Maribor, Brežice, Slovenia

P. Dapprich
Department of Agriculture, Fachhochschule Südwestfalen University of Applied Sciences, Soest, Germany

F. de Graaf
Spectral, Amsterdam, The Netherlands

S. Goddek
Mathematical and Statistical Methods (Biometris), Wageningen University, Wageningen, The Netherlands

S. Milliken
School of Design, University of Greenwich, London, UK

G. Stouvenakers, S. Massart and M. H. Jijakli
Integrated and Urban Plant Pathology Laboratory, Université de Liège, Gembloux Agro-Bio Tech, Gembloux, Belgium

P. Dapprich
Department of Agriculture, Fachhochschule Südwestfalen University of Applied Sciences, Soest, Germany

F. de Graaf
Spectral, Amsterdam, The Netherlands

S. Goddek
Mathematical and Statistical Methods (Biometris), Wageningen University, Wageningen, The Netherlands

S. Milliken
School of Design, University of Greenwich, London, UK

H. Stander
Department of Animal Sciences (Division of Aquaculture), University of Stellenbosch, Stellenbosch, South Africa

G. Proksch and A. Ianchenko
Department of Architecture, College of Built Environments, University of Washington, Seattle, WA, USA

B. Kotzen
School of Design, University of Greenwich, London, UK

www.ingramcontent.com/pod-product-compliance
Lightning Source LLC
LaVergne TN
LVHW051046220926
848485LV00001BA/26